SILSOE, BEDFORD, ENGLAND Tel. SILSOE (0525)
MK45 4DT 60428

THE APPLICATION OF ECONOMIC TECHNIQUES
IN ENVIRONMENTAL IMPACT ASSESSMENT

Environment & Management

VOLUME 4

The titles published in this series are listed at the end of this volume.

THE APPLICATION OF ECONOMIC TECHNIQUES IN ENVIRONMENTAL IMPACT ASSESSMENT

by

David James
Ecoservices Pty. Ltd.,
Whale Beach, NSW, Australia

KLUWER ACADEMIC PUBLISHERS
DORDRECHT / BOSTON / LONDON

A C.I.P. Catalogue record for this book is available from the Library of Congress.

ISBN 0-7923-2721-7

Published by Kluwer Academic Publishers,
P.O. Box 17, 3300 AA Dordrecht, The Netherlands.

Kluwer Academic Publishers incorporates
the publishing programmes of
D. Reidel, Martinus Nijhoff, Dr W. Junk and MTP Press.

Sold and distributed in the U.S.A. and Canada
by Kluwer Academic Publishers,
101 Philip Drive, Norwell, MA 02061, U.S.A.

In all other countries, sold and distributed
by Kluwer Academic Publishers Group,
P.O. Box 322, 3300 AH Dordrecht, The Netherlands.

Printed on acid-free paper

Printed in the Netherlands

TABLE OF CONTENTS

Foreword

The contribution of economic thought and method to environmental management needs practical illustration. Too few books on the subject achieve such an outcome. This book is among the notable exceptions.

That economics can provide a powerful vehicle for communicating an integrated understanding of the often diverse scientific findings germane to environmental impact assessment needs to be illustrated convincingly. This book does just that.

But it does more. It speaks across cultures: not to transfer know-how from one culture to another, but rather to activate an effective exchange of insights from one locale on the planet to another. As such, it is a genuine contribution to the great environmental exhortation of our times - think globally, act locally.

Too often the people best placed to make such contributions are too committed to practical outcomes and making a living doing so. Just occasionally, however, they can be persuaded to make the special effort required to communicate globally.

In this book, David James has once again orchestrated the contributions of virtuoso performers. In doing so he has emulated the contribution he sustained throughout the International Drylands Project and preparation of the books written with John Dixon and Paul Sherman: *The Economics of Dryland Management* and *Case Studies in Dryland Management* (Earthscan, London).

Taken together with his recent work as Special Commissioner for the path-breaking national Forest and Timber Inquiry for the Australian Government, we have a body of work characterised by great worthiness, integrity and true global significance.

DON MACRAE
CSIRO Corporate Planner

Preface and Acknowledgements

This volume is based on a number of reports prepared over several years for a research project jointly sponsored by the United Nations Environment Programme, the Australian International Development Assistance Bureau and the Australian and New Zealand Environment and Conservation Council. The author wishes to thank the three organisations for generous financial assistance to support the necessary research, and for their permission and encouragement to seek publication of the work.

Since the release of the report of the World Commission on Environment and Development, the quest has intensified for methods of analysis that combine economic analysis and environmental assessments. Within the discipline of economics, techniques of project and programme appraisal have reached a well developed stage, including approaches to the valuation of environmental effects. Environmental impact assessment, on the other hand, has been developed mainly by the scientific community and environmental agencies within government. Achieving the goals of sustainable development makes it imperative to blend these two traditions.

The main purpose of this volume is to explain how this may be accomplished. It attempts to combine scientific assessments, economic analysis and governmental procedures in an integrated approach to planning for economic development and environmental protection.

The first four chapters focus on concepts, techniques and administrative procedures. Six carefully selected case studies then illustrate integrated assessments in six different countries. Each case study contains a description of the relevant legislative and administrative framework and requirements for EIA, provided by Ben Boer and Angus Martyn from the Faculty of Law, University of Sydney. The author greatly appreciates their informative and valuable contributions.

Special thanks go to Don MacRae, previously Director of the Research and Information Branch of the Australian Department of Arts, Sport, the Environment,

Tourism and Territories, who played a catalytic role in initiating the project and who gave continued moral support throughout.

The author expresses his profound gratitude to co-authors and contributors to the volume: for their commitment to the overall concept, their diligence in preparing material and their patience in waiting for the volume to be brought to fruition. Without their efforts, the volume would not have been possible.

A significant intellectual debt is owed to my colleagues at the (former) Environment and Policy Institute at the East West Center in Honolulu, the Institute for Environmental Studies at the Free University of Amsterdam, and the research team that conducted the energy-economic-environmental modelling work for the Australian Government in the early 1980s, funded by the Australian National Energy Research, Development and Demonstration Council.

Richard Carpenter, from the East West Center, was particularly cooperative in helping to find appropriate case studies for the project. Blair Bower, John Dixon and Huib Jansen read early versions of the first four chapters, and made many useful suggestions. Brett Odgers, Project Officer for the report prepared for the Australian and New Zealand Environment and Conservation Council, gave invaluable guidance and personal support. Naturally, none of these people can be held responsible for any errors, omissions or deficiencies that may remain in this volume.

Permissions were granted by various publishers to reproduce materials over which copyright is held. Citations are appropriately documented in the volume. The author wishes to thank:

The American Chemical Society (*Environmental Science and Technology*)
Johns Hopkins University Press
Organisation for Economic Cooperation and Development
Science and Technology Letters (*The Environmentalist*)
Springer-Verlag (*Environmental Management*)

Personal thanks are extended to John Dixon for his permission to draw on materials from a study on Bacuit Bay and for reviewing that part of the text.

Further thanks are due to Tomago Aluminium Company Pty Ltd for allowing the author to reproduce material from the EIS prepared for their proposed venture in 1980 by J.B. Croft and Associates. Where material has been reproduced from the EIS, or commentary made on the EIS, the relevant source is acknowledged. The author wishes to make clear that any commentary made on material extracted from the report is not attributable to Tomago Aluminium Company Pty Ltd or the Participants in the venture.

A big thankyou to Mieke. She won't let me say any more.

Final thanks go to Mr Kisuk Cheung, Director of Engineering, Pacific Ocean Division, US Army Corps of Engineers, who gave consent to cite material from the Detailed Project Report and Environmental Impact Statement for Kahawainui Stream, on which Chapter 9 is based. Mr Cheung requested particularly that acknowledgement be given to the staff of the Pacific Ocean Division, that conducted the assessment, and of whom he is rightly proud. The author would like to take the liberty of quoting a paragraph from Mr Cheung's letter, which so aptly captures the spirit that many of us share:

"It is clear that the global environment and economy are the greatest challenges that we face in this modern world. I support our role in addressing these issues with enthusiasm and vigor. I look forward to our continued contribution and leadership in these exciting endeavors."

DAVID JAMES
Whale Beach
May 1993

LIST OF CONTRIBUTORS

Professor *Ben Boer* is the Corrs Chambers Westgarth Professor of Environmental Law and Director, Australian Centre for Environmental Law, Faculty of Law, University of Sydney. He has written extensively and has acted as consultant on environmental and resource law in relation to Australia and the Asia-Pacific region, including collaborating in the drafting of the proposed *Environment Protection Act* in Nepal.

Dr *Neil Byron* is Director of the Program for Graduate Studies in Environmental Management for Development at the National Centre for Development Studies at Australian National University. At the time of preparation of his case study, he was Senior Lecturer in Resource Economics in the Department of Forestry at Australian National University.

Alison Gilbert is a Researcher in the Institute for Environmental Studies, Free University of Amsterdam. In 1982 and 1983 she was a member of the team that worked on the NERDDC Project: An Integrated Energy-Economic-Environmental Model for Australia.

Geoff Goldrick is Lecturer in Geography in the Department of Geography and Environmental Science at Monash University. He has worked on a number of environmental consultancies with Ecoservices Pty Ltd.

Professor *David Griffin* is Deputy Vice-Chancellor of Australian National University. Previously he was Professor of Forestry at Australian National University and Director of the Nepal-Australia Forestry Project.

Dr *Tor Hundloe* is Environment Commissioner at the Industry Commission, Commonwealth Government of Australia. His case study was prepared while he was Director of the Institute of Applied Environmental Research, Griffith University.

Dr *David James* is a consultant and Director of Ecoservices Pty Ltd in Sydney. Positions previously held include Research Fellow at the Environment and Policy Institute, East West Center, Hawaii; Visiting Scholar at the International Federation of Institutes for Advanced Study in Maastricht; and the Rotating Chair in

Environmental Science at the Institute for Environmental Studies, Free University of Amsterdam.

Angus Martyn is a Research Associate in the Centre for Environmental Law in the Faculty of Law, University of Sydney. His work is primarily in the areas of environmental impact assessment and sustainable development. He is currently engaged in a major project on environmental law in Indonesia.

Chapter 1

CONCEPTS AND APPROACH

This volume is intended to assist all participants involved in environmental impact assessment (EIA): public and private development proponents, environmental and conservation groups, public administrators and decisionmakers, and interested members of the public. The focus is on the application of economic concepts and techniques.

EIA is a generic term that embraces both an administrative process and a set of analytical techniques designed to predict and appraise environmental impacts of development proposals. Such proposals may be put forward by the private and public sectors.

Economic analysis is becoming increasingly important as a planning and evaluation method in environmental assessments. The importance of undertaking a comprehensive economic analysis at the earliest possible stage of development proposals cannot be overemphasised. The main advantages are more effective screening and re-design of proposals before it is too late or costly to change direction.

The economic approach offers a logical means of integrating applied science and public decisionmaking, of reducing conflicts in environmental and natural resource management, and of reaching balanced decisions on development and environmental protection. Newer concepts in economics may be applied to the somewhat elusive goal of achieving "sustainable development".

A strong advantage of economics in EIA is that it provides a framework for the collection, analysis and interpretation of information. Strictly speaking, the main purpose of an "economic analysis" is to assess the efficiency with which resources, including the environment, are used to meet community preferences. With recent advances in valuation methods, it has now become possible to place economic values on many - but not all - environmental impacts and incorporate them in benefit-cost appraisals.

D. James, The Application of Economic Techniques in Environmental Impact Assessment, 1–32.

Although the main focus of an economic analysis is efficiency of resource use, techniques may be applied to other matters of concern in an EIA, such as the financial feasibility of proposed developments, secondary impacts on income and employment, and fiscal impacts.

This volume deals with the application of economics in EIA from technical and procedural viewpoints. The approach has been to follow the generic administrative procedure of EIA, indicating the role of economic analysis at each stage.

Development-environment decisionmaking can be divided into various steps, each the responsibility of a particular government authority. In the "model" presented here, development proposals are received, screened and authorised by a "Determining Authority" otherwise described as a Consent, Authorising or Approval Authority. If an EIA process is required, subsequent administration of the process is managed by an "Environment Authority" or Planning Authority. After public comment and in-house review, conducted jointly with the proponent and other government agencies, the Environment Authority conveys the results and recommendations to the Determining Authority, which makes a final decision and issues directives to the proponent. (These terms are used by federal and state governments in Australia and New Zealand. Their equivalents exist in other countries.)

Outline of the Book

In this chapter, general economic concepts are presented. EIA procedure is described in Chapter 2. Techniques of economic analysis are explained in Chapters 3 and 4. Six case studies are then presented, demonstrating practical application of the concepts and techniques under different sets of environmental conditions and in different parts of the world - Indonesia, Thailand, Nepal, the Philippines, Hawaii and Australia. Reference to the case studies is made, as appropriate, in the chapters presenting and explaining the various economic concepts and techniques.

Development and Environment

The Development-Environment Relationship

Perceptions of the relationship between environment and economic development have changed significantly over the last two decades. Under the old approach, development was seen to be a beneficial process, while impacts on the environment were considered to be mainly detrimental. Thus the typical decision problem for public authorities was to assess the tolerable cost of environmental damage in order to achieve the economic benefits of development. A common solution has been to proceed with the original proposal and minimise environmental damage through appropriate project design and management.

The new view of environment has a stronger economic basis. The environment is now seen as a set of resources - predominantly natural resources - that constitute part of the nation's capital stock. Goods and services yielded by natural capital have recognisable economic value to the community, despite the possibility that many such goods and services are not traded in markets and are not valued explicitly in terms of market prices.

Development involves a transformation of capital stocks to produce goods and services that are demanded by the community. Developments may be proposed as individual projects, as sectoral or regional development plans or as natural resource management plans. Previously, the emphasis in development was on the formation of man-made capital. Attention is now being focused on the role and economic value of natural capital.

"Sustainable development" recognises a relationship between economic development and environmental protection that takes into account the productive role of natural resources and the environment. Uses of the environment may include passive recreation and aesthetic enjoyment as well as the maintenance of life support systems.

As all economic development depends either directly or indirectly on the available supplies and quality of natural resources, stocks of natural capital should be preserved, conserved, expanded, rehabilitated, restored or improved to support long-term economic development. If natural capital is not properly managed, the prospects for long-term economic development may be limited.

Failures of the Market System

One explanation for the failure of markets in decisions concerning use of the environment is related to the prevailing system of property or use rights. Resources owned by nobody or everybody are incorrectly considered to be "free". Instead of their use being rationed through user costs, free resources are overused, often resulting in their degradation or destruction.

The other common explanation of market failure is "external" economic effects, which take place when the actions of one economic party lead to costs (or benefits) for some other party, without compensation of costs (or appropriation of benefits) because of the legal or institutional structure of society and the economy. Use rights and external effects are complementary concepts. They both have a common legal or institutional basis.

The Role of EIA

The EIA process, especially if combined with economic analysis, offers one set of public administrative procedures within which the goals of sustainable development may be pursued. EIA is a process of ensuring that proper consideration is given to environmental effects in development decisions. EIA procedures vary between countries and even between different jurisdictions within countries. Variations relate mainly to the degree of administrative discretion, public review provisions, legal standing to appeal against government actions or to initiate the EIA process, inquiry mechanisms, and links with land use planning and pollution control regulations.

The main application of EIA is in development projects, although EIAs may also be conducted for broad policies, general development plans, programmes and resource management plans. All of the case studies presented in this volume have regional environmental implications.

EIA provides an opportunity to modify or reject development proposals because of predicted environmental damage, and to consider alternatives that are capable of meeting the goals of the proposal.

EIA also provides a forum for participation in development decisions by a broad spectrum of participants, ranging from the proponent to the general community.

Participation in Development-Environment Decisions

Developments are typically initiated by a developer or proponent - a private individual, private company, public company, private consortium or government enterprise. The range of participants subsequently involved in development-environment decisions depends to a large extent on the type of development and on legislative or administrative requirements for environmental assessment. Some developments do not require an EIA; some require an EIA automatically; and others are at the discretion of a Determining Authority. In some instances, a Minister or other political head may call for an EIA.

The first point of contact between the proponent and government occurs when the proponent submits a preliminary development plan or notice of intention to the Determining Authority. The Determining Authority is responsible for initial screening of the plan and for issuing directives for further planning, including an EIA if warranted.

Administration of the EIA is usually the responsibility of an Environment Authority or Planning Authority. This Authority coordinates the assessment process, including referral to relevant government agencies.

A multidisciplinary team is needed to carry out the EIA. Most commonly, this consists of a team of professional consultants hired by the proponent. Some government agencies may conduct their EIA using in-house resources.

Following the environmental assessment, a draft environmental impact statement is prepared for the proponent and is made available for public comment. Simpler forms of EIA may not be subject to public review.

After public comment, a final environmental report is prepared. Clearance of the final report is given by the Environment Authority. Approval to proceed with the proposal is given by the Determining Authority. In some jurisdictions, there may be a right of appeal through legal processes.

Documentation of the Planning Process

The first report required in the development process is a preliminary planning document or notice of intention. Initial screening by the Determining Authority, or observance of automatic requirements, will determine whether an EIA should be prepared. If an EIA is required, the proponent is reponsible for the relevant planning document. The usual form of this document is an Environmental Impact Statement (EIS).

The guidelines for preparation of an EIS laid down by the Commonwealth Government of Australia provide an example of the required contents of an EIS. The EIS should:

- state the objectives of the proposed action;
- contain a description of the proposed action;
- indicate the consequences of not taking the proposed action;
- examine any prudent and feasible alternative to the proposed action
- include information and technical data adequate to permit a careful assessment of the impact on the environment of the proposed action;
- describe the environment that is likely to be affected by the proposed action and by any feasible and prudent alternative to the proposed action;
- assess the potential impact on the environment of the proposed action and any feasible and prudent alternative to the proposed action, including, in particular, the primary, secondary, short-term, long-term, adverse and

beneficial effects on the environment of the proposed action and of any feasible and prudent alternative to the proposed action;
- outline the reasons for the choice of the proposed action;
- describe and assess the effectiveness of any safeguards or standards for the protection of the environment intended to be adopted or applied in respect of the proposed action;
- cite any sources of information relied upon and outline any consultations during the preparation of the EIS.

Preparation of the EIS usually proceeds in two stages:
- preparation of a draft EIS, which is then subject to public comment and review by the Environment Authority;
- preparation of the final EIS.

Economics and EIA

Aims of Economics and EIA

Economics and EIA both aim at ensuring that all the effects of development, both within and outside development proposals, are taken into account. The "best" balance of development and environmental protection is assessed from the viewpoint of the community as a whole. The incidence of adverse and beneficial effects on particular groups or regions in the community may also be important in final decisions.

Fundamentally, economic analysis and EIA are applied as a means of "internalising" environmental effects. This process involves placing values on damage to or improvements in the environment as part of development planning. By placing economic values on goods and services yielded by the environment, the value of natural assets should be recognised more readily. The environment is then less likely to be damaged as an unintended or unaccountable consequence of development. The application of economics in EIA offers a practical mechanism for protection of the environment, including the design of mitigating measures.

The main difference between traditional EIA and an economic approach to EIA is that, under traditional EIA, judgements about the "significance" of effect tend to be derived in an ad hoc manner, whereas under the economic approach an attempt is made to measure rigorously social costs and benefits and express them in a familiar unit of measurement: monetary value (James et al. 1978).

Although economic techniques are capable of valuing a wide range of environmental impacts, it would be presumptuous to claim that all effects can, in practice, be valued in monetary terms. Sometimes, there is insufficient scientific information on environmental effects. Other effects, such as the valuation of human life, may rely more on ethical judgements than economic measurement.

This does not mean that "immeasurables" are ignored in an economic analysis. They are included in the balance sheet of benefits and costs and may be assessed in a descriptive way together with other more easily measured economic values. Where uncertainty prevails, economics can help to determine the most effective use of scarce research resources to address critical information gaps in the assessment. Indeed, the whole purpose of economic appraisal is to reduce uncertainty about the effects of proposed development.

Special constraints may be introduced for targets relating to human life or ecosystems. To the extent that "immeasurables" affect the outcomes of decisions, implicit valuations will be made.

Presentation of Results of Economic Analysis

The results of economic analysis undertaken within an EIA may be documented in two different ways:

- The EIS may contain separate sections dealing with the main social, ecological and engineering aspects of the development proposal, which are summarised and interpreted in a separate section on the economics of the proposal;

or

- the entire EIA may be conducted as a benefit-cost analysis, in which case the social, ecological and engineering aspects will be subsumed by the economic analysis.

The former approach is probably the most practicable, since the various specialists participating in an environmental investigation are usually proficient in preparing information for a particular field of analysis. The main role of economics is synthesis, evaluation and provision of recommendations on the ranking of alternatives and design options.

One of the advantages of an economic analysis is that the results may be presented in different degrees of detail, such as summary tables of benefits and costs (with a description of special problems or features of the proposal) for use at higher levels of government, and more detailed expositions in which all major assumptions and conclusions are explained, for use by experts, government officers and the general community.

Legislative Requirements for Economics in EIA

Legislative and administrative requirements for EIA vary considerably between countries. This is evident in the six case studies presented in this volume, for which the political, legislative and administrative settings are described. Different definitions of environment may be used. Some countries define the environment in physical and biological terms. Others include social and economic impacts.

The broader definitions of environment permit or encourage the incorporation of economic analysis in EIA. Even where narrower definitions apply, there is no special reason why physical environmental impacts should not be evaluated in economic terms.

Theoretical Foundations and Value Judgements of Benefit-Cost Analysis

Pareto Potential Welfare Criterion

Benefit-cost analysis has its origins in welfare economics. Standard references on benefit-cost analysis and project appraisal include Gittinger (1982), Little and Mirrlees (1976), Mishan (1982), O'Riordan and Sewell (1981), Pearce (1978, 1983), Pearce and Nash (1981), Squire and van der Tak (1975) and the Treasury Board of Canada (1978).

A change in resource allocation is considered desirable from the viewpoint of society as a whole if at least one person gains and all others are made no worse off. This rarely occurs in practice; thus another criterion is generally used, known as the Pareto potential welfare criterion, which deems a change in resource allocation desirable if the gainers are, in principle, able to compensate the losers and be no worse off. Benefit-cost analysis is based on the Pareto potential welfare criterion.

The underlying value judgements of benefit-cost analysis are:

1) Individual preferences count.
2) Individual preferences should be weighted according to the existing distribution of income.
3) Social preferences can be determined by summing the weighted preferences of all individuals comprising society.

Economic Welfare of the Individual

Economic valuation of benefits and costs begins with the values or preferences of individuals, as revealed in their economic behaviour. Individuals are assumed to be the best judge of their own welfare. Economic welfare of the individual is usually defined in terms of the "utility" that is derived in consuming goods and services. Individuals are assumed to spend their income on goods and services so as to maximise their utility. Utility can also be derived from the consumption of goods and services that are provided free.

In Figure 1.1 utility of the individual is indicated on the vertical axis and the consumption of some commodity X per unit of time on the horizontal axis. The total utility received by the individual is shown by the curve TU. As increasing quantities of X are consumed, total utility increases, but at a diminishing rate. At saturation point, so much is consumed that total utility begins to decline.

The utility that an individual receives by consuming an extra unit of the commodity is known as marginal utility, shown as the curve MU in Figure 1.1. Marginal utility is initially high but decreases as more is consumed. Beyond the saturation point, marginal utility is negative. In graphical terms, marginal utility is measured by the slope of the total utility curve. Algebraically, marginal utility consists of the derivative of the total utility function.

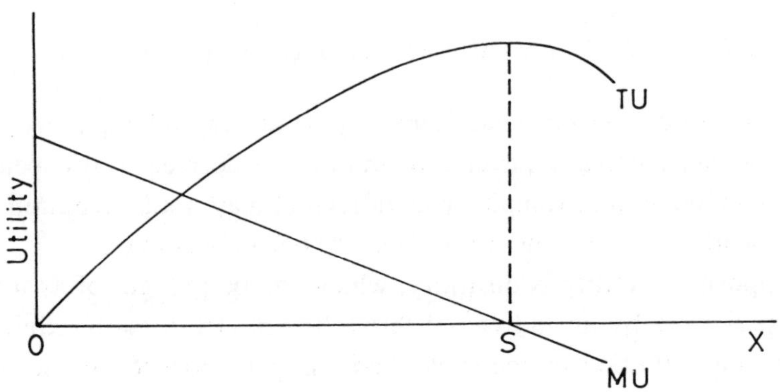

Figure 1.1. Total and marginal utility for good X.

The value the individual places on each additional unit of X will be reflected in the price he/she is prepared to pay. The marginal utility curve, in conjunction with available income, gives rise to a demand curve for the individual, shown as dd' in Figure 1.2. Typically, the demand curve is downward sloping, reflecting a decreasing willingness to pay for each marginal unit as an increasing quantity is consumed. At a given price (op) the individual will demand the quantity oq.

The total benefit obtained by the individual from consuming oq of X is measured by the area odrq. This area represents the total amount of money the individual would have been willing to pay if forced to pay separately for each unit consumed from o to q. The area consists of the amount of money the individual would actually spend on X, equal to the area oprq, and the area of the triangle dpr, known as consumer's surplus. The total willingness to pay thus consists of the individual's cash outlay plus consumer's surplus.

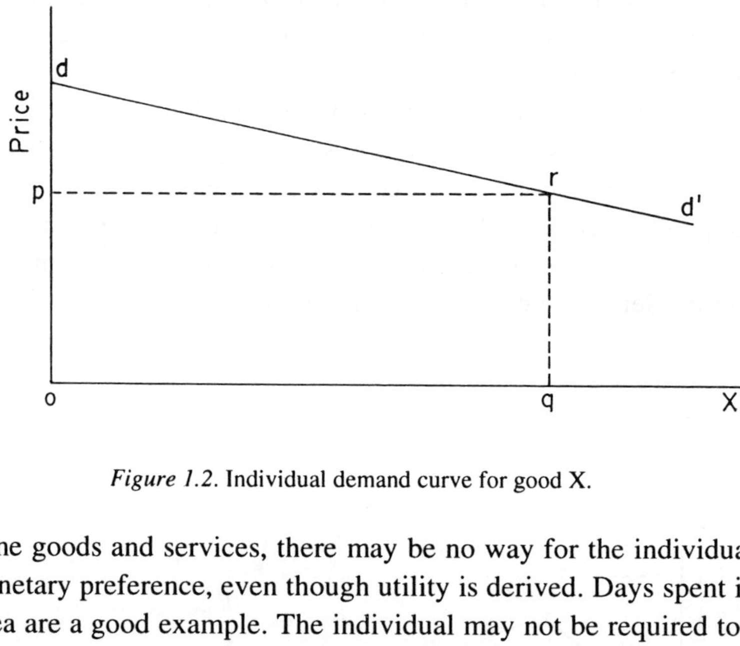

Figure 1.2. Individual demand curve for good X.

For some goods and services, there may be no way for the individual to express a monetary preference, even though utility is derived. Days spent in a wilderness area are a good example. The individual may not be required to pay for use of the wilderness area, but utility is nevertheless generated.

The opposite of utility is disutility, which can be thought of as the negative utility experienced by an individual through exposure to some kind of "bad". Environmental pollution is one such "bad". Deprivation of the use of an environmental amenity, as would occur if a wilderness area were destroyed, is another possible cause. Disutility may be expressed by an individual as a willingness to accept monetary compensation for the "bad" or his/her willingness to pay for avoidance of the effect. Monetary measures of disutility are classified as costs in an economic analysis.

In assessing the economic significance of the effects of development proposals, there is a reciprocal relationship between benefits and costs. A benefit is a cost avoided, and a cost is a benefit forgone. These concepts are important as a basis for classifying the effects of development into the broad categories of benefits and costs, for the purpose of economic measurement, evaluation and policy decisions.

Social Economic Welfare

Social benefits from consuming any good X are measured in terms of a market demand curve, which is derived by horizontally aggregating the demand curves for good X for all individuals in society. This is shown as DD' in Figure 1.3. Economic costs of resources used in producing and supplying good X are measured as an opportunity cost, or the economic value of the same resources in their next best alternative use. The costs of production are reflected in a market supply surve (SS' in Figure 1.3) which shows the price required by producers to cover the cost of producing each additional unit of good X. Any excess of revenue over cost earned by producers is known as producers' surplus. Where the market supply and demand curves for good X intersect, an equilibrium price (OP) and quantity (OQ) will be obtained. The marginal valuation of good X by all consumers will correspond to the marginal opportunity cost of resources used in production.

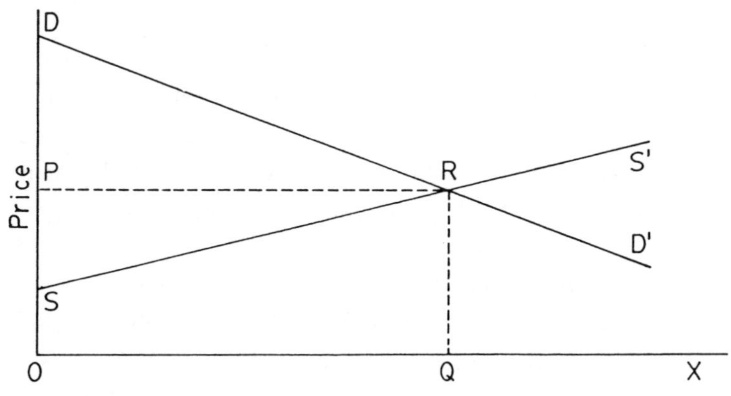

Figure 1.3. Market supply and demand for good X.

Eonomists use this analytical framework to measure the direct benefits and costs of development projects. Total social benefits are represented by the area ODRQ, consisting of the cash actually paid by consumers in the market (area OPRQ), plus consumers' surplus (area DRP). Total costs, assuming no fixed costs, are represented by the area OSRQ. Net social benefits are maximised when the market supply and demand curves intersect. In Figure 1.3 net social benefits consist of consumers' surplus (area DRP) and producers' surplus (area SPR).

As with individuals, society as a whole may experience utility or disutility from effects occurring outside normal market mechanisms. Monetary expressions of such values, for example of the enjoyment of wilderness areas or of damage caused by air pollution, can be found in the community's willingness to pay or to accept compensation, as the case may be.

Measures of changes in costs and benefits over time can be made comparable by expressing them in present values or as an annualised figure. This involves use of a social discount rate, which is similar to the rate of interest, but which reflects society's time preference for the consumption of goods and services, or the productivity of capital. The time horizon of the analysis should be established.

Prices and values in benefit-cost analysis are always expressed in real terms, that is, excluding the effects of inflation. The analysis may allow for changes in relative real prices, but not for changes in the general price level. The discount rate should also be expressed in real terms. Market rates of interest usually contain an inflationary component which should be corrected in a benefit-cost analysis.

Two other features of benefit-cost analysis should be noted, both of which underpin much of the analysis in the six case studies. First, it is important to identify the spatial boundaries of analysis, within which environmental effects may occur. The analysis is usually divided into an assessment of the "on-site" effects of a develoment proposal; and the "off-site" effects. What economists refer to as "environmental externalities" usually comprise off-site physical and other impacts.

Second, it is essential to establish the "with-project" and "without-project" situations. In establishing the without-project situation, the present baseline conditions should initially be identified. However, it must be realised that these conditions may not continue unchanged in the future, even if the project were not undertaken; the baseline itself may shift over time. In defining the without-project situation, prediction of future conditions, both economic and environmental, is thus necessary. Separate predictions are required of future conditions if the project were actually undertaken (the with-project situation). Benefit-cost analysis focuses on the net changes that are predicted to occur; that is, the differences between the without-project and with-project situations.

Decision Criteria in Benefit-Cost Analysis

Applications of Benefit-Cost Analysis

A benefit-cost analysis of a development proposal permits the relevant government agency to determine whether, in the broad sense, the proposal would use available resources (including environmental goods and services) efficiently from a community standpoint. It may not be the only criterion taken into account; it is possible that other considerations may over-rule the goal of economic efficiency. Nevertheless, the results of a benefit-cost analysis do indicate the preferences of the community in relation to the use of resources and the pattern of development.

Important applications of benefit-cost analysis include: helping to decide whether to accept or reject a development proposal; how to rank alternatives; and how to design and manage development to achieve a desirable balance of ordinary goods and services demanded by the community, and appropriate protection of the environment. For a particular development proposal, relevant questions to be answered in conjunction with the economic analysis are:

- Should the proposed development be undertaken or not?
- What is the scale of the proposed development?
- What is the best design?
- How should the proposal be modified to take environmental protection goals into account?

The NPV Criterion

The general rule for making these decisions is:

$$\text{maximise NPV} = B_d - C_d - C_p + B_e - C_e + B_o - C_o$$

where NPV is the net present value of the development alternative, B_d and C_d are the present values of direct benefits and costs, C_p is the present value of environmental protection costs, B_e represents the present value of environmental benefits, C_e is the present value of remaining environmental damage, and B_o and C_o are the present values of other indirect or secondary benefits and costs.

NPV is calculated by means of the formula:

$$\text{NPV} = \sum_{t=1}^{n} \frac{(B_t - C_t)}{(1+r)^t}$$

where B_t are economic benefits at time t, C_t are economic costs at time t, and r is the social discount rate.

If the NPV of the proposal is positive, the proposal should be accepted, but if it is negative, it should be rejected. Choice of the discount rate affects NPV. Usually, the higher the rate, the lower is NPV. An example of the calculation of NPV in a benefit-cost analysis is given in Table 1.1.

Table 1.1

Estimate of NPV of Ranger Uranium Project ($A million in January 1976 prices)

Year June	Total Benefits	Capital Expend.	Operating Expend.	Net Benefits	Discount Factor	Present Value
1977	0.0	13.5	0	-13.5	1.000	-13.5
1978	0.0	32.2	0	-32.2	0.909	-29.3
1979	0.0	49.7	0	-49.7	0.826	-41.1
1980	9.4	22.3	12.8	-25.7	0.751	-19.3
1981	58.7	2.4	26.5	29.8	0.683	20.4
1982	97.2	0.7	27.3	69.2	0.621	43.0
1983	98.4	1.5	27.3	69.6	0.564	39.3
1984	99.6	3.0	27.3	69.3	0.513	35.6
1985	99.8	2.2	27.3	70.3	0.466	32.8
1986	102.4	5.8	27.4	69.2	0.424	29.3
1987	104.8	4.7	27.4	72.7	0.385	28.0
1988	104.8	1.0	27.4	76.4	0.350	26.7
1989	104.8	2.4	27.4	75.0	0.319	23.9
1990	104.8	3.6	27.4	73.8	0.290	21.4
					NPV =	197.2

Assumptions:

Full production = 3,000 tonnes per annum

Price of yellowcake = US$20 per pound

Exchange rate US$1.26 = $A1.00

Discount rate = 10 per cent

Source: Commonwealth of Australia (1976)

Assessment of Alternatives

EIA legislation and regulations usually require that alternatives to a development proposal be assessed. Alternatives may be appraised through benefit-cost analysis and ranked according to their NPV. The alternative with the highest NPV is the most desirable in terms of the efficiency of resource use. This rule assumes that there are no absolute constraints over resource availability, such as investment funds. As long as NPV is positive, it will be economic for society to allocate resources to the project.

Internal Rate of Return

It is often suggested that development proposals should be ranked according to their internal rate of return (the discount rate that equates the present value of benefits with the present value of costs). However, ambiguous assessments are often derived. The criterion that should be applied is NPV.

Benefit-Cost Ratio

The benefit-cost ratio (the ratio of the present value of benefits to the present value of costs) is another evaluation criterion. It is obvious that if the ratio exceeds 1 a proposal should be accepted, since this is equivalent to a positive NPV; conversely, if the ratio is less than 1 the proposal should be rejected. However, as with the internal rate of return, benefit-cost ratios should not be used to rank alternatives.

Capital Rationing in Public Agencies

In public investment programmes, use of the NPV criterion must be modified where there is rationing of capital by the investing agency. This topic, however, is beyond the scope of this book. The reader is referred to a relevant text such as Lumby (1984), Levy and Sarnat (1978) or Quirin and Wigington (1981) for the appropriate ranking criteria.

Broader Applications of Benefit-Cost Analysis

Benefit-cost analysis can be used to appraise broader development policies, regional development plans and natural resource management strategies. Simulation models are frequently used to explore alternative scenarios, which are designed, evaluated and ranked according to their benefits and costs.

Financial Analysis

Discounted Cash Flow Analysis

Financial appraisal of projects is necessary to determine whether the proposal is commercially viable. For public sector investments, financial appraisals provide the basis for funding and pricing arrangements. Unless financing aspects are satisfactory, development proposals cannot go ahead, regardless of their economic value to the community.

The modern approach to financial appraisal is based on discounted cash flow analysis (DCF). It deals with the pecuniary costs and benefits faced by the developer. The accepted method is to calculate the net present cash value of the proposal. The formula for the present value of net cash flows is

$$PV = \sum_{t=1}^{n} \frac{(R_t - C_t)}{(1+i)^t}$$

where PV is the present value of the net cash flows, t is each time period, R_t is gross revenue at time t, C_t is total costs at time t, and i is the rate of interest. The number of time periods in the life of the project is n.

If a project involves some kind of negative net benefit initially, as would occur in a capital installation phase, funds borrowed or invested in the project will be costed at the market interest rate. As the project breaks even, the rate which the funds could earn if the surplus were invested is again the market interest rate.

Differences Between Financial and Economic Analysis

Commercial profitability is not always a good indicator of social profitability. When the social profitability of a development activity is to be evaluated, benefit-cost analysis should be undertaken. The main differences between financial analysis and economic analysis concern the way in which costs and benefits are defined, the prices used to value the costs and benefits and the treatment of costs and benefits over time (see Table 1.2).

In contrast to the costs and benefits in a financial analysis, economic costs and benefits are not defined narrowly in terms of a development proponent's cash flow. The accounting stance used in financial analysis is that of the development proponent, whereas in economic analysis the accounting is done from the point of view of society as a whole. Costs and benefits which are "external"

Table 1.2

Differences Between Financial and Economic Analysis

Characteristics	Financial Analysis	Economic Analysis
Purpose	Indication of incentive to adopt	Determine whether investment is justified in economic efficiency terms
Accounting Stance	Developer	Society
Discount Rate	Marginal cost of borrowing money in financial markets	Social discount rate
Stance on Transfer Payments	Relevant to analysis	Not relevant to analysis
Prices	Market prices	Shadow prices may be required
Costs	Price of all inputs	Opportunity cost of all inputs
Benefits	Revenues	Real benefits to the economy as a whole

to the developer's financial calculations, such as the costs of off-site environmental degradation, are "internalised" in an economic analysis.

Shadow Prices

A financial analysis values inputs and outputs in terms of market prices. However, prices in an economic analysis may not correspond to the prices that rule in the market. In a benefit-cost analysis, prices are measured in terms of the community's marginal willingness to pay (WTP) or in terms of the marginal opportunity cost of resources. "Opportunity cost" can be defined as the value of resources in their next best alternative use.

Market prices frequently do not reflect marginal willingness to pay or marginal social opportunity cost. Where this occurs, "shadow prices" which correct for distortions in the economy are used in the economic analysis. For instance, a financial analysis uses market prices for factors of production such as land, labour and capital. When unemployment of these factors exists, economic analysis uses a shadow price to reflect their true opportunity cost. Where unemployment in the labour market prevails, the opportunity cost of labour will usually be much less than the going wage rate. The case studies on Thailand and Nepal illustrate this point.

Where market prices do not exist at all, as is often the case for environmental benefits and costs, the willingness to pay or opportunity cost of inputs and outputs must be imputed by means of surrogate markets or contingent valuation methods, which are explained in Chapter 3. When changes occur in the availability of a good, large enough to affect market prices, consumers' surplus should be calculated and included in the analysis.

Interest Rate and Discount Rate

The market interest rate used in financial analysis may not reflect the rate at which society discounts future costs and benefits. In economic analysis, a discount rate based on the social opportunity cost of capital and/or the social rate of time preference is used. The discount rate used in financial analysis may be either nominal (incorporating an inflationary component) or real (adjusted for inflation), while in economic analysis the discount rate is always the real rate.

Converting Financial Values to Economic Values

Despite the conceptual differences outlined above, the information used in financial analysis can often be used as a basis for economic assessments. "Corrections" are made to financial data to convert them to economic estimates for inclusion in an economic analysis. Some of the more important corrections are described below.

Economic analysis uses real, inflation free price estimates of costs and benefits. Where future costs and benefits are expressed in nominal prices, these should be converted by means of a price index to the relevant base year, usually the starting year of the project. Inflation can generally be ignored except where it is expected to proceed at uneven rates in different sectors of the economy. In this case, care must be exercised in predicting the correct relative prices of goods and services. All changes in real relative prices over time must be accommodated in the economic analysis.

From the point of view of economic efficiency analysis, transfer payments such as taxes and subsidies do not represent allocative costs or benefits. Transfers do not alter the total production possibilities and consumption opportunities in the economy; they merely constitute a change in the distribution of income. The prices used in economic analysis should therefore be net of taxes and subsidies.

Depreciation allowances are a device used by accountants to write off capital equipment for tax purposes. Depreciation estimates or other forms of amortization have no place in economic analysis. In economic analysis capital costs consist of the opportunity costs of the resources used to create the capital stock. The real costs of capital formation are incurred when the resources are actually committed. Sunk costs should be ignored as they do not represent an opportunity cost of the development activity.

Benefits and Costs of Environmental Protection

Context of Environmental Mitigation

Mitigating the adverse environmental impacts of development proposals is an important objective of EIA procedure. This requires assessments of the alternatives that are available within proposed projects to meet environmental quality goals. Part of the assessment and design phase is the determination of the level of protection that should be achieved, how it should be measured and evaluated, and the extent to which costs can or should be borne to bring about the required level of protection (Freeman 1979,1982; Kneese 1984).

Basically two types of assessments must be made:

1) Fulfilment of design standards required by the Environment Authority.

Legislated standards or regulations often apply to the control of potential environmental impacts at or within the physical boundaries of a project. Controls over effluent discharges or emissions are a good example. Other controls may be applied to natural resource development projects. For instance, in forestry operations, regulations may preclude logging on steep slopes, to minimise soil erosion and protect catchments.

Prescribed standards must be met by a proponent regardless of cost. An attempt should be made, however, to find the least-cost solution. Economic analysis, particularly cost-effectiveness analysis (see Chapter 3), can assist in this task. If the cost of meeting prescribed standards is too high, the proposal may need to be abandoned. This ideally should be dealt with in the early stages of the proposal, to avoid unnecessary subsequent investigation.

2) Determination of optimal tradeoffs for environmental effects not covered by existing standards.

There may be no existing rulings on the control of adverse effects on, for example, natural ecosystems or rare species. In addition, even where existing standards are met, serious environmental damage may occur because of site-specific or region-specific environmental characteristics. The damages caused by pollution cannot always be controlled by legislated emission standards. Damage is associated with ambient concentrations rather than emissions as such. The connection between emission rates and ambient concentrations is often complicated. Dispersion processes and chemical reactions in the environment must be taken into account.

Negotiation between the proponent and the Environment Authority may be required to determine the "best" type and level of control. Project assessors and designers can make initial suggestions in the preliminary EIS, but subsequent approval or modification by the Environment Authority will be needed. A tradeoff must be made between the costs of achieving higher levels of environmental protection and the benefits of control (damage costs avoided). The Environment Authority is responsible for ensuring that some degree of economic efficiency is attained in reaching an acceptable design.

The benefits of environmental protection accrue largely off-site, to society as a whole. The costs of mitigation must be met by the proponent, but society itself ultimately has to pay (Australian Environment Council 1983). Costs can usually be passed on to buyers of the goods or services provided by the proponent. If too high a level of protection is imposed and costs cannot be passed on, the proposal may not be financially feasible. Compensation, subsidies or other fiscal support can be provided by the government if it is decided that the proposal must proceed and certain levels of environmental protection are to be achieved.

Economic Models of Environmental Protection

The economic approach to environmental protection and the design of projects to mitigate effects looks at tradeoffs between the costs of protection measures and the environmental benefits that can be achieved. The optimal level of protection maximises the net benefits of resource use (including the environment as a resource) or equivalently, minimises the costs of protection and damage to the environment (Hufschmidt et al. 1983).

This approach is based on the concept that it is economically irrational, beyond a certain point, to allocate an exorbitant amount of resources to the control of environmental damage, and equally irrational to allow significant environmental damage to occur from development proposals. Where predicted environmental damage is extreme, as in cases involving acute toxic effects or irreversible damage to unique ecosystems, it may be desirable to implement stringent and possibly expensive environmental protection standards rather than rely on economic assessments.

Mitigating the Impacts of Development Projects

The economic principle can be demonstrated for the control of emissions from a development project, for example, an electric power station. To assess the economic implications of different levels of emissions from the project, it is necessary to construct an environmental damage cost curve. This is shown as the curve DC in Figure 1.4. The greater the level of emissions, the higher the cost of damage. Note that the environment has some assimilative capacity; OD emissions can be discharged without causing any damage. Assume that the predicted level of emissions from the project is OP. The predicted level of economic damage of the project would thus be PC. Emissions from the project can be reduced by control measures, but only at increasing cost. The abatement cost curve PB rises from right to left.

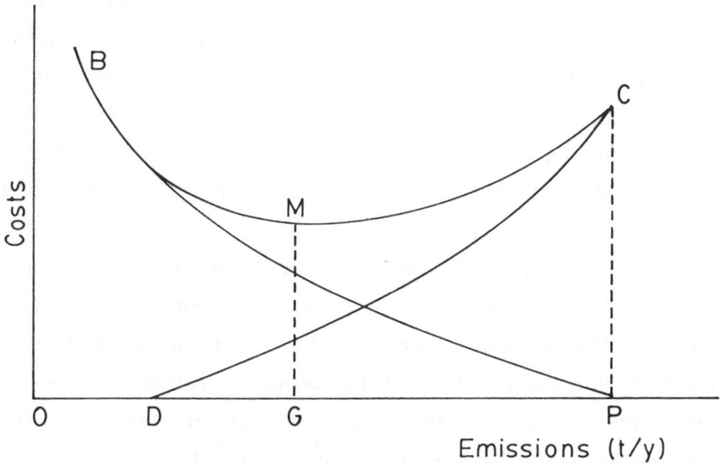

Figure 1.4. Damage costs and abatement costs for emissions..

Vertical addition of the two curves gives a U-shaped curve BMC indicating the total cost of emissions. The minimum point on this curve is at M. The diagram shows that the total costs of emission control are minimised when the level of abatement is PG and the acceptable level of emissions is OG. These conditions occur where the marginal abatement cost equals the marginal damage cost.

To understand why this approach is equivalent to maximising the net benefits of emissions control, refer to the formula for NPV (p. 13), where C_p and C_e indicate environmental protection costs and environmental damage costs respectively. Maximising $(-C_p-C_e)$ in the NPV formula is equivalent to minimising (C_p+C_e) as in the example here. Damage costs avoided are a benefit of the project. Other environmental benefits B_e could also be created by the project, such as a large artificial lake for cooling purposes, that may be used for swimming, boating or as a habitat for native birds.

The principle demonstrated here for a development project may also be applied to broader policies and programmes of environmental protection addressing environmental concerns expressed by the community.

Handling Risk and Uncertainty

Problems of Risk and Uncertainty in EIA

Risk and uncertainty are endemic in all environmental impact assessments and in economic appraisals of development proposals. Economic valuation takes place at two points in the assessment methodology: first in predicting and assessing direct inputs and outputs of proposed development (in the "with" and "without" situations) and their valuation in terms of costs and benefits; and the prediction and valuation of "external effects". The economic valuation of environmental effects depends on prior predictions of the physical, social and ecological impacts of proposed development in the "with" and "without" situations. Such predictions must be made by specialists other than economists.

An economic assessment is thus exposed to all the uncertainties of the usual scientific, sociological, engineering and ecological impact assessments as well as the uncertainties of its own predictive, analytical and valuation methods. Special consideration must be given to the results of an economic assessment and ways of addressing the problem in providing information to support decisions on development and the environment.

Definitions of Risk and Uncertainty

In economic analysis there is a difference between risk and uncertainty. Risk may be defined as the probability and magnitude of an adverse event. Where the event can vary in magnitude, the risk of its occurrence must be defined in terms of a probability distribution. Random or "stochastic" variables are commonly encountered in environmental assessments. Relevant examples include flooding, occurrence of wild species and pollution caused by industrial accidents. Many economic variables are also random, such as movements in the market prices of goods and services, shifts in technology and production costs and changes in community preferences and values. Relevant risks in an economic analysis lead to a probability distribution for the net benefits of a proposed development.

It is worth noting that, where a probability distribution can be specified for net benefits, an analysis of risk need not be confined to the downside or adverse prospects of a proposed development; over a particular range of the distribution there may be prospects of above-average net benefits. For reasons that will be subsequently explained, the community and hence public decisionmakers tend to focus on the risk of adverse effects.

With uncertainty, the probability distribution is unknown, even for upper and lower bounds of effect, so the likely extent of adverse impacts cannot be quantified. Further, the pathways and mechanisms leading to effects may be poorly understood or largely unknown. The greenhouse effect is an example of environmental impact where considerable uncertainty prevails.

Management of Risk

Risk (and uncertainty) may be handled through risk management, which has three elements: risk assessment; risk appraisal or evaluation; and response to risk. These are discussed in greater detail below.

Risk Assessment. Risk assessment is the objective assessment of effects and their corresponding probabilities. The expected value of an effect is calculated as the magnitude of the effect multiplied by its probability. Where there is a range of effect, the expected value is the mean (arithmetic average) of the distribution.

Risk Appraisal. Risk appraisal attempts to assess the community's attitude towards various kinds of risk and, if possible, to quantify the value that is placed on risk. Social valuations of risk can be presumed to depend on individual valuations of risk.

A benchmark for measuring the risk preference of an individual is the "certainty equivalent" which is the guaranteed benefit that would make the individual as well off as with a risky benefit. Where an individual's certainty equivalent corresponds with the mathematical expected value, he/she is said to be risk neutral. If the certainty equivalent is less, the individual will be risk averse. He/she will be willing to pay a premium, sometimes referred to as an "option value", to eliminate risk. Insurance premiums paid to counteract the risk of medical costs or motor accidents are relevant examples. Where the certainty equivalent exceeds the expected value, the individual is said to be a risk taker.

The community tends to be risk averse in relation to adverse enviromental impacts. Perceived risks are often much greater than expert assessment would suggest. For example, the mere existence of certain toxic substances in the environment, however small the concentrations and regardless of prevailing background levels, may be enough to induce the community to reject some types of development proposals, such as pulp mills or sewage disposal plants.

There is abundant empirical evidence indicating that risk to which the community is involuntarily exposed, such as air pollution in cities, has a larger negative value than risk that is voluntarily undertaken, as in motor sports or scuba diving. The community is also generally risk averse for effects that may have catastrophic consequences, such as nuclear accidents or dam failures. Even though the expected value of damage from such events may be very small, the potential maximum magnitude of effect may lead to extreme risk aversion by the community. The possibility of irreversible environmental damage such as the extinction of species or ecosystems also tends to generate large negative values on the part of the community.

Community attitudes to risk, whether judged rational or irrational by public decisionmakers, can affect significantly the success or failure of development decisions. It is not always easy to measure the willingness of the community to forgo benefits or pay the price to reduce environmental risk, but the usual direction of preference - risk aversion - may be presumed.

This leads to the conclusion that the net benefits of development projects and programmes with high levels of risk, particularly of damage to the environment, should be adjusted downwards. It is not recommended that a "risky discount rate" be applied. Risk should be assessed specifically in relation to the elements of the analysis for which probabilistic information is available.

There are, unfortunately, no clearly defined rules determining how far adjustments should be carried. It must be noted, nevertheless, that risk is a part of every decision and that relative risk is what matters, not the absolute level of risk.

Response to Risk. Where risk cannot be avoided, its effect on community welfare can be alleviated by two mechanisms: risk pooling and risk spreading. Risk pooling refers to the diversification of investments (developments) whereby the variance of possible net benefits and potentially adverse effects on the environment is reduced. Risk spreading is the dilution of adverse impact over large numbers of individuals in the community, so that the burden for each individual becomes tolerable. Motor vehicle insurance is an example of risk spreading.

In many cases, the response to risk may allow for risk prevention (lowering the probability of adverse impact) or risk reduction (lowering the magnitude of effect). Such benefits may be achieved through careful design of a project, engineering works and management plans for ecosystems at risk. An example of risk prevention is an improvement in the reliability of aircraft landing systems. Flood mitigation is an example of risk reduction. The case study in Chapter 9 applies economic risk analysis to flood damage assessment and the design of mitigation measures for a watershed on the Island of Oahu, Hawaii.

Dealing with Uncertainty

Uncertainty can be handled in various ways in an economic analysis. What should first be ascertained is the elements of the evaluation to which uncertainty applies. Whether uncertainty is a major problem could depend on whether these elements are a significant component of benefits or costs. It may be necessary to focus on the timing of uncertain impacts, especially in conjunction with the discounting of future costs and benefits. Other solutions include using subjective judgments by experts as to the probabilities of effect, applying decision analysis techniques and conducting research to reduce the level of uncertainty.

Sensitivity Analysis. Sensitivity analysis is a simple procedure designed to assess the significance of uncertainty in an economic analyis. "Best case" and "worst case" values may be tried for uncertain values in the benefit-cost analysis. Different temporal patterns of impact and different discount rates may also be assumed. Evaluations such as accept/reject or the ranking of alternatives are often unaffected by changes in uncertain values in an economic analysis.

Subjective Probabilities. Uncertainty can be handled through the techniques of risk assessment and risk appraisal if probabilities for uncertain effects are prescribed. For example, medical experts may be asked their opinion on the likely health impacts of different levels of exposure to air pollutants or toxic substances in foodstuffs. Ecologists could express an opinion on the probabilities of extinction of plant or animal species as a consequence of changes in habitat. The Delphi technique is a useful way of obtaining expert assessments. Other methods include surveys and questionnaires.

Decision Analysis Techniques. Decision analysis techniques assess possible outcomes of actions (for example, development programmes and projects) under different conditions. The information may be represented by a decision tree, which depicts decision sequences and their consequences.

Another way of presenting information is in the form of a payoff matrix, which displays potential benefits or costs of alternative outcomes. The attitude of the decisionmaker towards risk allows the preferred alternative to be identified. A common decision criterion is risk aversion. Under this approach, the worst-case outcomes for each alternative are selected. The highest of these is then selected. This criterion is sometimes described as the "maximin" criterion.

Decisionmakers may instead choose to minimise regret. The regret associated with each alternative is an opportunity loss, measured as the difference between the payoff from an alternative that is arbitrarily chosen and the highest payoff that might have been achieved in retrospect. The minimum regret criterion identifies the alternative that results in the lowest opportunity loss.

Reducing Uncertainty by Research. If uncertainty is a major problem in reaching decisions on development approval, one strategy is to postpone the decision until more can be learned about the likely nature of effects. The cost of postponing action has been described as a "quasi-option" value. Clearly, there is a tradeoff between the costs of delay and the benefits of reduced uncertainty.

Under an approach known as adaptive environmental assessment and management (AEAM) it is recommended that project managers monitor the effects of their actions and adapt their behaviour as more is learned about the system under consideration. Such an approach may be applied to pollution control programmes with uncertain effects, such as the reduction of greenhouse gas emissions. A logical sequence is to conduct preliminary research, design remedial action, undertake action and research simultaneously, then discontinue research. It would be difficult to escape subjective assessments of the probabilities of costs and benefits of action and research under this approach.

References on Risk and Uncertainty

Useful references on risk and uncertainty include Environmental Resources Ltd (1985), Dixon et al. (1989), Norton (1984), Pearce and Nash (1981) and Pouliquen (1970). Guidelines on the estimation of expected flood damage are provided by ANCOLD (1986), Penning-Rowsell and Chatterton (1977) and Parker et al. (1987). Smith and Handmer (1986) discuss flood warning systems. The standard reference on adaptive environmental assessment and management is Holling (1978).

Sustainable Development

Brundtland Definition of Sustainable Development

Sustainable development has become a policy target for governments at all levels. Only the beginning steps have been taken to incorporate decision rules for sustainability in environmental impact assessments and economic appraisals. One of the difficulties in operationalising the concept has been a lack of agreement on a suitable definition. The Brundtland Commission (World Commission on Environment and Development 1987) defines sustainable development as "development which meets the needs of the present without compromising the ability of future generations to meet their own needs". The Brundtland definition is mainly concerned with keeping options open. In particular, it stresses the potential benefits of protecting natural resources and the environment. According to the Commission, economic development and environmental protection should be mutally reinforcing goals.

Rules for Sustainability

Sustainable development has proved to be an elusive concept in practice. Attempts to put the concept into operation may be facilitated by finding answers to the following questions:

- What is to be sustained?
- Within which set of spatial boundaries?
- Over what period of time?
- For whose benefit?
- Who will determine what is to be sustained?

Two rules suggested by Pearce et al. (1989) to implement sustainable development are:

1) Renewable resources should be managed sustainably.
2) Resource rents from exhaustible resources should be reinvested in resources to compensate for the exhaustible resource when it is depleted.

Ecologically Sustainable Development

The term "ecologically sustainable development" carries the sustainable development concept one step further. It emphasises the need to protect the ecosystem processes on which all human activity ultimately depends. In this sense, natural resources and ecosystems can be considered part of the community's capital stock (James et al. 1989).

Much of the current debate about ecologically sustainable development relates to differences in perceptions and assessments of how much natural capital should be preserved, in what form and how it should be managed (including, in some contexts, the possibilities of a "hands-off" management approach). Economically sustainable development allows for continual substitution between man-made and natural capital; it focuses on the capacity to sustain Gross Domestic Product (GDP). Ecologically sustainable development involves the growth of GDP together with the conservation or enhancement of natural capital stocks.

Those who advocate ecologically sustainable development contend that there are limitations to the extent to which natural capital can be replaced by man-made capital. If all the benefits of natural capital could be assessed properly, an acceptable balance might be defined, but in practice such an assessment is difficult to carry out. Often, the benefits can only be presumed. In addition, the tyranny of small decisions pervades many environmental assessments; each impact of itself may seem insignificant but the collective impact on ecosystems as a whole may involve serious risks of damage. A holistic impact assessment methodology is thus advocated as the most effective means of identifying the total impacts of development.

Discount Rates and Sustainable Development

Discount rates reflect two economic factors. First is the time preference of the community. Empirical evidence indicates that most people value something they can consume now more than when it is consumed in the future. To persuade people to abstain from consumption now in favour of the future, a premium is required. In financial markets, this is the rate of interest. The rate used in an economic analysis is the social rate of discount. It may be likened to a real rate of interest. Second is the rate of return on capital or the opportunity cost of capital. Resources diverted to capital formation or development projects should yield a net return equivalent to the return that may be earned by investing the same resources elsewhere in the economy. The discount rate reflects this rate of return.

The discount rate is thus a measure of the intertemporal efficiency of resource use. It should be selected with care. Many public agencies prescribe an official discount rate.

Discount rates operate in reverse to compound interest. The higher the discount rate, the lower will be the present value of future benefits and costs. After forty or fifty years, the present value of future benefits and costs is very small. Planning horizons for economic analysis should thus be chosen in conjunction with the discount rate.

Arguments Against Discounting

It is frequently contended that use of a discount rate in economic analysis is incompatible with the concept of sustainable development. The reasons cited for this view are that discounting disregards future costs and benefits and hence ignores future generations. It is thus tempting, so the argument goes, to bequeath degraded environments and their costs to future generations; further, there will be little incentive to undertake many environmental protection or improvement programmes and projects because the benefits occur too far into the future; both of these actions will have little effect on calculations of NPV.

Another argument against the use of discounting in the management of renewable resources is that, if the natural rate of growth of the natural stock is less than the discount rate, economic analysis would suggest appropriating the stock and investing the proceeds elsewhere in the economy. Discount rates, that is, encourage resource extinction.

Use of a zero discount rate is thus sometimes advocated in the interests of environmental protection and ecologically sustainable development.

Discounting and Environmental Impacts

Despite objections to the practice of discounting, especially in relation to environmental effects, there are sound reasons for its use. First, use of a zero or lowered discount rate raises practical difficulties, in that assessments would be required of impacts to an infinite planning horizon. Second, decisions on accept/reject and on the ranking of alternatives are not always sensitive to the discount rate. Often, the same conclusions hold true in an economic assessment regardless of the discount rate that is applied. Much depends on the time paths of net benefits of the various options. Third, many environmental protection measures, programmes and projects yield benefits almost immediately. The alleviation of beach pollution and the introduction of flood mitigation works are relevant examples.

Discount Rates and Resource Extinction

It is true that, in the management of renewable natural resources such as fisheries and forestry, application of a discount rate will tend to reduce the size of the optimal stock. It is also true that if the natural rate of growth of renewable stocks is less than the discount rate, it may appear "economically rational" to harvest the stock and invest the benefits in other areas of economic activity. But if the resource is sufficiently productive, sustainable use of the stock may still be compatible with the application of a discount rate. Much depends on the inititial size of the stock, growth rates and costs and returns. Commercial forestry plantations, managed on a sustainable basis, have expanded rapidly in recent years with quite acceptable commercial rates of return. Evidence of this is provided by the case study in Chapter 7.

Even where extinction of the stock is indicated, it must be realised that changes in the composition of the economy's capital stock, including natural capital and man-made capital, are a normal part of the development process and may be necessary to generate higher levels of community economic welfare - for example, the conversion of some forested land to agriculture.

Nevertheless, where extinction of a natural capital stock is suggested by economic analysis, it is prudent to assess whether all the costs and benefits of predicted changes in the stock have been properly measured. For example, significant external benefits may be associated with the stock, such as "existence values" for whales or amelioration of the greenhouse effect for forests. Mostly, concern over resource extinction is associated with the prospect of irreversible impacts.

Irreversible Environmental Damage

The risk of irreversible damage to the environment, without prior understanding of benefits forgone, is of major policy significance. The prescription to assess development options more carefully, particularly in terms of environmental impacts, is merely a reflection of the precautionary principle. Although irreversible environmental damage is based on physical and ecological considerations, it is at the same time an economic concept. The degree of irreversibility depends on the cost of restoration, rehabilitation or substitution. Some environmental impacts can be reversed at modest cost, as in reafforestation programmes. Other impacts may be prohibitively expensive, as in the case of species extinction or the accumulation of persistent pollutants. Again, if the relevant costs and benefits could be assessed in full, it might be easier to reach decisions on development and resource use.

The connection between irreversible damage and ecologically sustainable development is the possibility that irreversible destruction of natural capital may preclude forever sustainable yields of environmental goods and services by the ecosystems affected. Many of these goods and services are not exchanged in markets, and their valuation is plagued by uncertainty.

In the absence of better information, application of the precautionary principle to ecosystem damage may be recommended. This does not necessarily mean that a "hands-off" approach to the environment is warranted. Safe minimum standards may be applied in development decisions and in resource management plans (Ciriacy–Wantrup 1952; Ciriacy–Wantrup and Bishop 1975; Goodland and Ledec 1987). By conducting research, the environmental systems under threat may be better understood and patterns of development may be guided along desired lines (Dasgupta 1982). Rational management of resources is not likely to be achieved by manipulating the discount rate in economic assessments.

Discount Rates and Intergenerational Equity

Social time preference is an ethical judgment. There is nothing to preclude ethical decisions about bequeathing certain natural (or other) assets to future generations, but adjusting the discount rate is not an efficient means of influencing decisions. Additional policy targets or constraints are a far more effective mechanism. It should also be noted that the preferences of future generations are unknown. Decisions on intergenerational equity must be based on what the current generation believes would contribute most to the welfare of future generations.

References

ANCOLD (1986) *Guidelines on Flood Design for Dams*, Australian National Committee on Large Dams.

Australian Environment Council (1983) *The Costs of Pollution Abatement: A Preliminary Survey of Nine Australian Industries*, Australian Government Printing Service, Canberra.

Ciriacy-Wantrup, S.V. (1952) *Resource Conservation Economics and Policies*, University of California Press, Berkeley and Los Angeles.

Ciriacy-Wantrup, S.V. and R. Bishop (1975) 'Common Property as a Concept in Natural Resources Policy', *Natural Resources Journal*, 15:713-737.

Commonwealth of Australia (1976) *Ranger Uranium Environmental Inquiry, First Report*, Australian Government Publishing Service, Canberra.

Dasgupta, P. (1982) *The Control of Resources*, Blackwell, Oxford.

Dixon, J.A., D.E. James and P.B. Sherman (1989) *The Economics of Dryland Management*, Earthscan, London.

Environmental Resources Ltd (1985) *Handling Uncertainty in Environmental Impact Assessment*, Ministry of Public Housing, Physical Planning and Environmental Protection, The Hague.

Freeman, A.M. (1979) *The Benefits of Environmental Improvement*, Johns Hopkins University Press, Baltimore.

Freeman, A.M. (1982) *Air and Water Pollution Control*, Wiley Interscience, New York.

Gittinger, J.P. (1982) *Economic Analysis of Agricultural Projects*, Johns Hopkins University Press, Baltimore.

Goodland, R. and G. Ledec (1987) 'Neoclassical Economics and Principles of Sustainable Development', *Ecological Modeling*, 38:19-46.

Holling, C.S. (ed) (1978) *Adaptive Environmental Assessment and Management*, Wiley, Chichester.

Hufschmidt, M.M., D.E. James, A.D. Meister, B.T. Bower and J.A. Dixon (1983) *Environment, Natural Systems and Development: An Economic Valuation Guide*, Johns Hopkins University Press, Baltimore.

James, D.E., H.M.A. Jansen and J.B. Opschoor (1978) *Economic Approaches to Environmental Problems*, Elsevier Scientific, Amsterdam.

James, D.E., P. Nijkamp and J.B. Opschoor (1989) 'Ecological Sustainability and Economic Development' in F. Archibugi and P. Nijkamp (eds) *Economy and Ecology: Towards Sustainable Development*, Kluwer, Dordrecht.

Levy, H. and M. Sarnat (1978) *Capital Investment and Financial Decisions*, Prentice-Hall, Englewood Cliffs.

Little, I.M.D. and J.A. Mirrlees (1976) *Project Appraisal and Planning for Developing Countries*, Heinemann, London.

Lumby, S. (1984) *Investment Appraisal*, Van Nostrand Reinhold, London.

Mishan E.J. (1982) *Cost Benefit Analysis*, Praeger, New York.

Norton, G.A. (1984) *Resource Economics*, Edward Arnold, London.

O'Riordan, T. and W.R.D. Sewell (eds) (1981) *Project Appraisal and Policy Review*, Wiley, Chichester.

Parker, D.J., C.H. Green and P.M. Thompson (1987) *Urban Flood Protection Benefits: A Project Appraisal Guide*, Gower, Aldershot.

Pearce, D.W. (1978) *The Valuation of Social Cost*, Macmillan, London.

Pearce, D.W. (1983) *Cost Benefit Analysis*, Macmillan, London.

Pearce, D.W., A. Markandya and E.D. Barbier (1989) *Blueprint for a Green Economy*, Earthscan, London.

Pearce, D.W. and C.A. Nash (1981) *The Social Appraisal of Projects: A Text in Cost Benefit Analysis*, Macmillan, London.

Penning-Rowsell, E.C. and J.B. Chatterton (1977) *The Benefits of Flood Alleviation*, Saxon House, Farnborough.

Pouliquen, L.Y. (1970) *Risk Analysis in Project Appraisal*, Johns Hopkins University Press, Baltimore.

Quirin, G.D. and J.C. Wigington (1981) *Analyzing Capital Expenditures: Private and Public Perspectives*, Irwin, Hornewood Ill.

Smith, D.I. and J.W. Handmer (1986) *Flood Warning in Australia*, Centre for Resource and Environmental Studies, ANU, Canberra.

Squire, L. and H.G. van der Tak (1975) *Economic Analysis of Projects*, Johns Hopkins University Press, Baltimore.

Treasury Board Canada (1978) *Benefit-Cost Analysis Guide*, Ministry of Supply and Services, Hull, Quebec.

World Commission on Environment and Development (1987) *Our Common Future*, Oxford University Press, Oxford.

Chapter 2

EIA PROCEDURE

There are many useful references on environmental impact assessment, which include Ahmad and Sammy (1985), Basta and Bower (1982), Biswas and Geping (1987), Black (1981), Canter (1977), Clark et al. (1980), Erickson (1979), FEARO (1985), Jain et al. (1977), Munn (1979), Nichols and Hyman (1980), O'Riordan and Hey (1976), and Rau and Wooten (1980). This chapter describes the procedure of EIA as it is widely practised in many countries.

EIA procedure should begin early in the planning and evaluation of projects and programmes and continue throughout the life of the activity as a management operation. In this chapter, the main steps of EIA are outlined. Then follows a more detailed discussion of the role of economics within each step. An explanation is given of the ways economics can be integrated with traditional assessments, the techniques that can be applied and incorporation of the results of analysis within planning documents.

Steps in EIA Procedure

In the "model" of EIA procedure presented here, the main steps of analysis are as follows (see Figure 2.1):

1) The proponent formulates the proposal. Prior to submission, the proposal may undergo revision to reduce adverse environmental impacts.
2) The proponent submits the proposal to the Determining Authority.
3) The Determining Authority screens the proposal to assess the significance of the likely environmental impacts of the activity. The significance of the activity will depend on its size, nature, kinds of environmental impact, ecosystems or human systems under threat and the likely magnitude of effects. To assist in this task, checklists or screening guidelines may be used.
4) The Determining Authority considers the scope and programme for an EIA.
5) The Determining Authority notifies the Environment Authority.

D. James, The Application of Economic Techniques in Environmental Impact Assessment, 33–61.

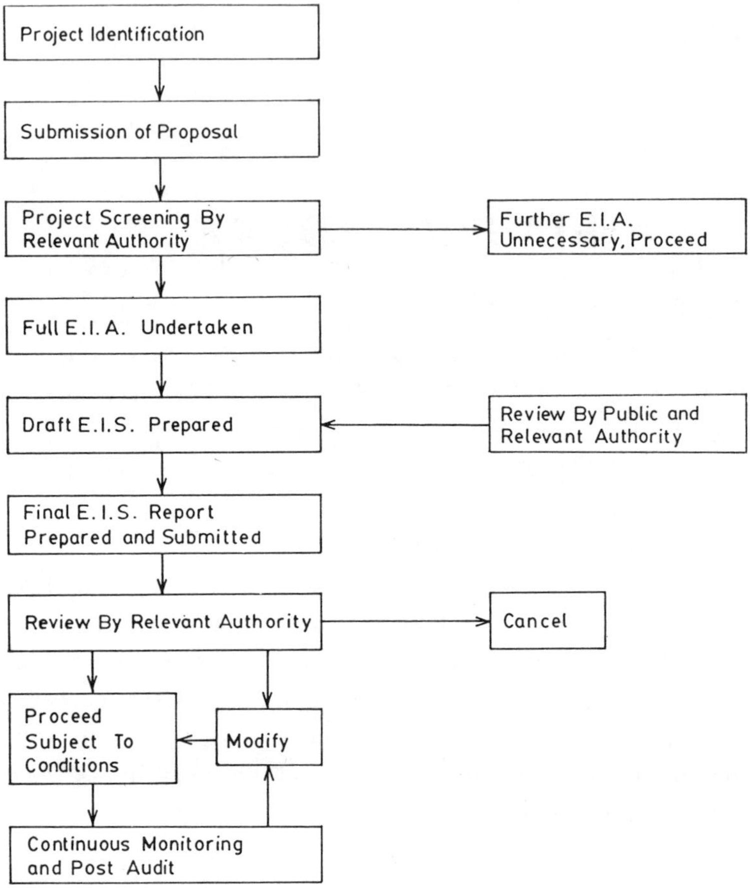

Figure 2.1. Environmental Impact Assessment procedure.

6) The Environment Authority determines the appropriate level of EIA, scopes the key issues, and establishes the terms of reference or guidelines for the EIA and any public review.
7) A detailed EIA is undertaken by the proponent, Determining Authority or competent authority and documented in an Environmental Impact Statement (EIS) or other planning document. The EIS should contain:
   - a description of the proposed activity and its need or rationale;
   - a description of feasible alternatives;
   - a description of the potentially affected environments;

- an assessment of all the likely or potential environmental impacts (including cumulative, direct, indirect, short and long term effects) of the proposed activity and alternatives;
- an identification, description and assessment of mitigation measures for each alternative;
- an account of provisions for post-audit.

8) A public inquiry or review may be held to enable citizens to comment on the contents of the EIS. Useful comments should be incorporated.
9) The EIS and the proposed mitigation and monitoring programme are reviewed by the Environment Authority.
10) The final EIS is prepared, including a full account of mitigation and monitoring measures.
11) The Determining Authority makes the decision to proceed or cancel.
12) A post-audit is undertaken by the Determining Authority and the Environment Authority to determine the accuracy of the predictions and to see whether mitigation measures were properly carried out.

Formulation of Development Proposal by Proponent

Development Planning and EIA

Development planning takes place in the private and public sectors as a routine operation. Existing environmental regulations or standards must be observed as normal requirements of planned developments. Whether the formulation of development proposals should extend to the broader - and more public - process of EIA depends on legislative or administrative provisions. A common prerequisite is that a proposal should have "significant" environmental effects.

The likelihood (or certainty) of an EIA being required should signal the importance of taking environmental factors into account at a very early stage of development proposals. Indeed, the need for an EIA may be prompted by preliminary indications of economic impacts, in the broadest sense.

EIA is not intended to be a one-off event occurring at a particular point in the planning process. Rather, EIA should be an integral part of development planning and management, from initial conceptualisation and design to the continuous monitoring of operations and eventual reinstatement of a site.

Private Sector Proposals

Where a proponent has reason to believe that an EIA will be required as part of the development consent procedure, there are distinct advantages in addressing environmental considerations at the formulation stage, thus avoiding possible delays and other complications at a later date. The general thrust of the development proposal may be influenced by environmental constraints or opportunities.

From a legal or administrative perspective, the EIA process potentially begins for a private proponent when an application for approval is made to the Determining Authority. By this stage, the proponent should have considered environmental requirements that must be reflected in the development application. The proponent would presumably have already carried out a detailed financial analysis and have judged this aspect of the proposal to be satisfactory.

Public Sector Proposals

For public sector development proposals, there may or may not be a legal requirement to consider environmental matters in the formulation stage. In many instances, public authorities are able to conduct their own environmental assessments. As in the case of private sector proposals, however, it will usually be prudent for public development authorities to allow for environmental factors at an early stage of the planning process, thereby minimising costly changes if objections are raised.

Public sector agencies are in a strong position to protect the environment as part of their development planning. Unlike private sector proponents, which are motivated primarily by financial returns, public agencies are (or should be) concerned with community welfare. Thus they are well placed, in the early phases of plan formulation, to apply design and evaluation criteria such as the maximisation of net social economic benefits, protection of species and ecosystems, and fulfilment of the goals of sustainable development.

Categorisation of Activities of Environmental Significance

Particular kinds of development are likely to have significant environmental effects. Some may automatically require an EIA under legislative or administrative requirements. Examples of environmentally significant developments are provided in the Environmental Screening Guidelines used by the Australian International Development Assistance Bureau (AIDAB 1989). They include the following:

- major construction and infrastructure projects including ports, airports, railways, roads, bridges, power lines and causeways;
- energy production including hydroelectric dams, thermal power stations, nuclear energy, oil extraction and processing;
- industrial development including petrochemical, processing of agricultural products, minerals processing, chemical production, manufacturing, pulp and paper;
- minerals extraction including all types of mining and quarrying, beach sand mining and dredging;
- agriculture and forestry particularly activities which involve land clearing, irrigation, introduction of new species, pest control and fertilizers;
- fisheries and hunting particularly if exploitation may affect sustainable harvest;
- integrated rural development;
- small-scale rural water and sanitation;
- improvement of basic urban facilities such as slum upgrading and transport planning;
- tourism development.

Characteristics of Proposed Development

Certain characteristics of proposed developments have the potential to affect the environment, which should be considered in the formulation phase. They include:

- products, by-products and/or services of the proposed development;
- process technology or production method;
- inputs to the activity;
- waste gases, liquids and solids, including toxic substances;
- energy, heat and noise;
- land requirements;
- exposure to risk;
- economic factors including employment and income generation;
- social and cultural factors;
- regional economic interactions;
- time profile for construction, operation and post-project management.

Initial Feasibility Assessment

Feasibility of the initial proposal must be assessed in terms of its financial viability, availability of materials, labour, managerial skills and other inputs, and operational efficiency. This is a usual task in project planning.

Meeting prescribed or anticipated environmental regulations, standards, constraints and safeguards can affect the nature and viability of initial proposals. Broad alternatives should be carefully assessed, including ways of meeting existing requirements for environmental protection.

After initial plan evaluations, a preferred plan may be selected by the proponent. This opens the way for a more thorough scoping of environmental effects. At this stage of the assessment procedure, however, broad-brush EIA combined with preliminary economic analysis could indicate cancellation or alteration of a proposal.

Scoping Assessment of Environmental Effects by Proponent

Description of the Environment

Scoping the assessment of environmental effects by the proponent should begin with a survey of the environment, both on-site and off-site. Relevant factors for the site survey include:

- physical boundaries;
- location;
- geology and soils;
- climate;
- water resources;
- fauna and flora;
- items of cultural or heritage significance.

For the off-site environment, similar factors should be taken into account. The social and economic environment should also be considered.

Baseline Studies

Baseline studies are required to establish initial environmental conditions. It is important to recognise that, in subsequent predictions of the likely environmental effects of the project, a comparison must be made between the "with-project" and "without project" situations. The "without project" situation refers to likely environmental conditions, including those taking place over time, in the absence of the proposed development, not just the set of environmental conditions prevailing prior to commencement of the project. Baseline studies should provide a framework for making predictions of environmental conditions over time, regardless of whether the development proposal is actually implemented.

Initial Environmental Effects

A screening study should next be undertaken by the proponent, identifying the most important environmental effects. Checklists are usually available from government agencies as part of the guidelines for environmental assessment procedures, indicating in general the kinds of environmental effects that might be important. It is necessary to identify only those effects relevant to the particular proposal.

The various phases of the proposal should be distinguished - for example, exploration and/or site survey, construction, operation and post-project management. A different set of environmental interactions is likely to be involved in each phase of the development.

Identification of the environmental effects of the proposal may be facilitated by using an interaction matrix, in which the various facets of the proposal are represented by the columns and the environmental effects by the rows. The main interaction cells may then be pinpointed. The matrix is an effective visual means of studying, in a qualitative way, the relationship between the proposed development and the environment.

It is not advisable to apply an arbitrary rating system to assess the significance of impacts, for example as under the Leopold matrix approach, as this introduces confusing and unnecessary detail. More importantly, if an economic approach is to be taken in the EIA, the relevant environmental impact values will be derived using the techniques of benefit-cost analysis, not the ad hoc procedures implicit in the Leopold approach (Leopold et al. 1971).

Interaction matrices are not the only way of screening the proposal for environmental effects. Screening guidelines, such as those used by AIDAB, perform a similar function, as do more formal techniques such as the Sorensen step matrix.

There are several good reasons for identifying, even in a rough qualitative way, significant areas of likely environmental interaction:

- the initial screening may reveal environmental effects that are unacceptable, in which case the proposal may have to be redesigned, other alternatives considered or the proposal abandoned;

- attention will be focused on particular problem areas requiring detailed data collection; and
- the screening should suggest the kinds of research skills and analytical techniques that will be needed to carry out the full assessment.

Boundaries must be established in the screening process, both temporal and physical.

Temporal Boundaries of Effect

Time boundaries are established with selection of the "planning horizon". For many engineering projects, this constitutes the expected physical life of a structure or activity. An open-cut coal mine, for example, may have an expected life of 10 years. Coal-fired electric power stations may operate for 50 years, with a major refurbish at 30 years. Hydroelectric schemes and reservoirs may have an expected life of 150 years or more. The economics of sustainable development recommends that if significant long-term environmental effects are predicted, a long-term planning horizon may and should be used.

The time path of effects is also important - for example, the length of the construction phase and the normal operational phase. The "boomtown" effect can be significant for major projects in specific regions.

Physical Boundaries of Effect

Physical boundaries should be determined according to the transmission of environmental effects. In the case of pollution, for example, the boundaries may be defined as an airshed or watershed. Arbitrary judgements are often required in establishing physical boundaries. It may not be possible to predict the ultimate environmental effects of a development proposal. Research skills or equipment may not be available, data and predictive methods may be inadequate, and there may be insufficient time for a detailed investigation. It may also not be possible to distinguish the effects of the proposal from the effects of other activities, as in the case of photochemical smog, acid rain or the greenhouse effect.

Even where the effects can be predicted in spatial terms, their incidence may be insignificant and barely worth incorporating in the analysis - for example, a slight increase in ambient concentrations of particulates, within accepted ambient concentration standards.

Determination of the physical boundaries of effect at an early stage of plan formulation will facilitate the selection of research personnel, data systems, analytical approaches and time budget for a full environmental investigation.

Social and Economic Boundaries

Appropriate boundaries should be established for social and economic impacts of the proposal. Impacts on social groups include changes in demographic factors, effects on the demand for housing, disruption of community patterns, loss of recreation amenity, changes in lifestyle and increased pressures on community service facilities.

If the proposal is large relative to the regional economy, some kind of regional economic impact analysis in a subsequent EIA would be indicated. One consideration is involvement of the local work force in the development. The value of output or income of the proposed development as a percentage of total regional output or income is another.

Identification of Special Environmental Effects

Some environmental effects warrant special consideration and should be identified in the early screening process. Such effects may include:

- threats to rare species and special habitats;
- irreversible effects;
- cumulative environmental impacts;
- synergistic effects;
- delayed environmental impacts;
- environmental risks or hazards;
- destabilisation effects (economic and ecological);
- release of toxic wastes;
- threats to cultural artifacts and sites;
- potential loss of heritage structures or areas.

Further special effects will depend on the type of development proposal and the environment. Some of the effects described above create difficult problems of economic valuation. However, as explained in Chapter 1, the presence of "immeasurables" in the analysis should not preclude the use of an economic framework and approach.

Other effects involve additional assessments or decision criteria that must be considered in conjunction with, or as part of, the benefit-cost analysis. The problem of risk and uncertainty is one example. Although economists can assess risk in objective terms, policy decisions should take into account community attitudes and values towards risk. Uncertainty in economic analyses creates further policy concerns. These problems have already been alluded to in Chapter 1.

Environmental impacts that are irrreversible, cumulative, synergistic or persistent are a potential threat to the goal of sustainable use of natural resources and the environment. The question of which discount rate should be used is often raised in connection with these more complex types of environmental effect. Simple economic efficiency analysis is not sufficient, of itself, to handle many of these problems. Again, however, approaches have been established that directly address these problems in an economic analysis.

Preparation of Preliminary Planning Document

Functions of Preliminary Planning Document

After conceptualisation and environmental screening have been undertaken by the proponent, and a decision has been made to continue with the proposal, a preliminary planning document is prepared. This document is submitted to Determining Authority with an application for development approval. It precedes an EIA.

The preliminary planning document has three purposes:

1) It serves as a management tool for the proponent to guide the design, implementation and operation of the proposed development, rather like a provisional business plan.
2) It informs government of the proponent's intent to undertake a development and describes the initial plan.
3) It provides an opportunity for government to screen the proposal for possible significant environmental effects, and determine whether an EIA is required.

Relevant Planning Criteria

Certain planning criteria are of primary concern to the proponent, such as:

- availability of an acceptable site, materials and energy inputs, labour, management expertise and finance;
- a well-defined time schedule for project implementation;
- an acceptable marketing strategy (if appropriate);
- financial acceptability to the proponent.

Other criteria are of concern to government agencies, but must be satisfied by the proponent for the proposal to succeed. These include:

- observance of design regulations or special directives;

- incorporation of environmental safeguards as prescribed under statutory standards, licenses or direct regulations;
- compliance with administrative procedures;
- provision for payment of royalties, taxes, charges and fees;
- compatibility with broader development plans such as local or regional environmental plans.

Content of Preliminary Planning Document

The preliminary planning document should include brief descriptions of the following:

- specification of the development proposal with a discussion of feasible and prudent alternatives;
- description of the implementation and management plan;
- description of the existing environment;
- indication of potential environmental impacts;
- outline of the proposed environmental safeguards and mitigation measures;
- description of monitoring programme;
- identification of relevant statutory requirements;
- scope for modification of the proposal;
- issues of special concern.

Inclusion of the following economic information will enhance the usefulness of the document to the Determining Authority:

- financial profitability or feasibility;
- expenditure on environmental protection;
- off-site financial impacts;
- economic interpretation of the major on-site and off-site benefits and costs.

Anticipation of an EIA

If an EIA is anticipated, the proponent must be aware of a broad range of community concerns, and be prepared to allow for possible changes introduced through public involvement or by government agencies representing community interests. This includes, but is not restricted to, off-site physical environmental effects.

Some on-site physical effects may have community implications. For example, a mining project or logging programme may threaten some rare species, involving "existence" or "preservation" values that extend to the broader community. Heritage values, cultural values and wilderness values may also be affected by on-site activities.

Economic and financial factors that would otherwise remain confidential may also enter the public domain if an EIA is required, especially if a social benefit-cost analysis is undertaken as part of the EIA. Such factors include sources of equity, connections with other companies, sources and value of inputs to production, commercial opportunities, taxation and other fiscal considerations, and regional economic effects. The proponent usually has the right, however, for sensitive commercial information to be witheld.

With a growing emphasis by governments on the economic implications of environmental effects and environmental protection measures, the proponent is advised to prepare and categorise environmental information in a form that lends itself to economic evaluation within the EIA.

Assessment of Preliminary Planning Document by Determining Authority

Screening for Significant Environmental Effects

Review of the preliminary planning document provides the first opportunity for government to screen the proposal for significant environmental effects. The review is an interactive process between the proponent and the Determining Authority.

In the past, the significance of environmental effects has been judged somewhat arbitrarily by engineers and scientists engaged in assessments of physical environmental effects. Economic analysis can be used in the screening process to identify key areas of effect, to help decide whether an EIA is required and to scope the terms of reference for the EIA if an EIA is considered to be necessary.

Indeed, early EIA and comprehensive economic analysis is vital at this stage. It may indicate the need for additional information, a broader consideration of alternatives or re-specification of the proposal. In extreme cases, it may become obvious, because of significant adverse environmental or economic effects, that the proposal should be abandoned. It is better to take such a decision at an early stage, rather than persevere and waste valuable time and investigative resources.

Commercial interests would also be served more effectively by early notification of the unacceptability of particular proposals. Indecision and delay by government can disrupt investment plans, lower business confidence and discourage new development proposals. Serious public conflict is also a possibility if proposals with unacceptable consequences are allowed to proceed further in the assessment process.

Identification and Valuation of Environmental Effects

Environmental impacts of the proposed development must first be identified in physical terms. Adverse environmental effects should be classified as costs and improvements as benefits. Note that it is the net impacts that should be identified, or the difference between the "with" and "without" development scenarios.

The economic importance of various aspects of the proposal may be assessed using benefit-cost analysis. Such an analysis should begin with a balance sheet of the main categories of benefits and costs. Benefits and costs relevant to the environment include environmental protection costs, environmental benefits and remaining environmental damage costs.

In the screening process undertaken by the Determining Authority, only rough economic valuations may be required. Impacts measurable in market prices (for example, damage to crops or fisheries) are the easiest to quantify. Other effects may require the use of surrogate market techniques or preliminary surveys.

Where a proposal has few environmental benefits, the predicted occurrence of large damage costs may indicate the need for re-design of the proposal, a consideration of other alternatives or outright rejection.

Cumulative Environmental Impacts

Cumulative environmental impacts may take several forms. One is the accumulation of substances in the environment, such as heavy metals in bottom sediments of lakes and rivers. Accumulation may also occur in biota, with potentially adverse effects on ecosystems or human health. Bioaccumulation of pesticides in fish and birds can result in mortality or impaired function. Ingestion of toxics by humans - for example, heavy metals or organochlorines can have serious health effects such as Minamata disease or cancer.

Other cumulative impacts may take the form of indirect or "spillover" effects within a region, resulting from production linkages between the proposal and other activities in the region. For example, establishment of an aluminium smelter may lead to increased air pollution by power stations supplying the necessary electicity, or to increased water pollution by coal mines producing coal for the power stations. An example of these impacts is provided in Chapter 10.

Cumulative impacts may have a delayed effect. Nitrates used as fertilisers, for example, are known to pose a threat to human health by contaminating groundwater supplies. It may take 15 years or longer for nitrates to reach groundwater basins.

There may also be thresholds in cumulative impacts. For example, sustained yields may be maintained for fisheries provided the harvest rate does not exceed net production of biomass. An increase in the harvest rate may lead to extinction of the resource.

Cumulative impacts are often associated with random disturbances. For example, forestry projects may be devastated by bushfires. In semi-arid ecosystems, sustained levels of grazing may be practised for long periods of time, but drought can easily result in destruction of the resource.

The Determining Authority should be especially aware of cumulative environmental impacts and consequent threats to the sustainable use of natural resources. There are limits to what can be assessed by an individual proponent. In addition, an individual proponent cannot be held responsible for the potential impacts of all other proponents or existing activities within a region.

Impacts of Special Concern

The Determining Authority should ensure that the proposal does not have serious deleterious environmental impacts. The prospects of irreversible damage, loss of unique species and habitats, exposure to risk, etc. must be screened carefully.

Environmental Safeguards and Mitigation

Information on technologies, project design and management practices for the purpose of environmental protection, together with the associated costs, should be considered by the Determining Authority to assess whether sufficient care has been taken by the proponent to prevent or mitigate adverse environmental effects.

The proponent should realise that if a full EIA is required, an economic assessment will probably have to be carried out for the environmental protection aspects of the proposal. Financial as well as broader social economic factors may need to be taken into account.

Financial Aspects of Proposal

The financial feasibility of the proposal can be of community significance in several ways. With public enterprises, the proposal may be acceptable only if certain forms of government assistance or cross-subsidisation of customers are applied. The possible extent of such assistance is usually an issue of public concern.

In the private sector, financial failure can have detrimental social, economic and environmental consequences. For example, if large-scale forestry and woodchip operations turn out to be non-viable financially and are forced to close down, severe regional economic impacts may result. Taking another example, mine closures caused by bankruptcy may leave large areas of disturbed land, leading to dust emissions, toxic leachates and adverse aesthetic effects. In some countries, mining companies are required to pay a bond to the government to ensure that rehabilitation will take place, even in the event of economic failure.

Private financial information may be needed for certain types of government approval. For example, to obtain an export licence, government approval of the export price may be required. This could become a key issue in development proposals for commodities such as woodchips or minerals.

The Determining Authority may request information on export prices, profitability and tax revenue if it suspects that the proponent is practising transfer pricing. This involves an arrangement between a local developer which is a subsidiary of a parent company in another country, to export to the parent at a price below world average. Profits are effectively transferred to the parent company. Tax minimisation is often a motive for transfer pricing. The result is that the host country fails to appropriate the full value of its exports and economic returns on its natural resources.

The Determining Authority may be interested in the financial aspects of a proposal, either public or private, if it has low financial returns but high off-site economic returns. Environmental improvement programmes and projects undertaken as extensions of development projects may fall within this category. Tax incentives or subsidies may be paid to support environmental protection goals. This is one mechanism that governments can use to encourage environmentally sustainable development.

The preparation of financial data has another advantage: it often serves as the basis for a social benefit-cost analysis. This process of converting financial to economic values has already been explained in Chapter 1.

Secondary Economic Impacts

Technical and economic linkages between the proposed development and the regional economy are important determinants of secondary or indirect economic and environmental impacts. A check on regional economic multipliers for the relevant industry, if available, will give a preliminary indication of the likely indirect impacts of the proposal. The usual source of such multipliers is an input-output table for the region (see Chapter 4).

The Determining Authority should be interested in regional economic linkages, as they may have important implications for the distributional effects of the proposal. There may be further implications for broader development plans. For example, the proposal may indicate a need for major public investments in infrastructure such as transport systems, water, sewerage and government services. The general nature of a region may be transformed as a result of induced expansion of industry.

Income Distribution Effects

Income distribution effects resulting from a proposed development constitute a social impact that may be of significance in public decisions on the proposal. The living standards of some social groups, such as aged people on fixed incomes, may be affected if the proposal leads to local inflation.

Fiscal Impacts

Before a proposal can proceed, government agencies at local, state and federal levels may need to be satisfied about the effects on their revenues and outlays. Local governments in particular may be obliged to make large investments in supporting infrastructure and services.

The Determining Authority may wish to recommend possible arrangements for covering public sector expenditures. Companies proposing new coal mines in the Hunter Region of New South Wales, for example, are obliged to make a front-end payment to local government to meet the infrastructure costs. Other alternatives may involve block grants by State or Federal governments. The review process should consider relevant taxes, royalties and other charges.

Compatibility with Other Plans and Policies

Checking the compatibility of the proposal with other plans and policies is an important responsibility of the Determining Authority. Examples of factors against which compatibility should be checked include:

- zoning policies and preferred uses of the environment;
- local and regional environmental plans;
- state or provincial development plans;
- federal development and resource conservation policies;

- environmental standards and regulations;
- sustainability objectives (avoidance of environmental synergisms, irreversibilities and cumulative impacts);
- resource availability in the region;
- social and welfare policies.

Decision on Proposal

EIA legislation and administrative guidelines are not always clear about what matters should be taken into account in deciding whether a proposed development does have significant environmental effects. Preliminary economic analysis of environmental effects can help identify the main problem areas.

The Determining Authority may wish to carry out an indicative benefit-cost analysis to assess the net effect of the proposal on community welfare. This analysis should suggest areas requiring more detailed investigation.

Proposals that are considered desirable by the proponent may not be desirable for the general community. A financial analysis of the proposal, for example, may make the proposal very attractive for the proponent, whereas a social benefit-cost analysis may indicate its rejection on broader welfare grounds. The main differences between financial and economic analyis have been explained in Chapter 1.

It may not, of course, be possible to evaluate all environmental effects in economic terms. Some aspects, such as the value of life, raise awkward philosophical issues. In other cases the research resources may not be available. Some techniques, such as contingent valuation methods, may be fairly expensive to apply, and interpretation of the results may create some difficulty for public decisionmakers. Effects that are difficult to measure in economic terms should nevertheless be taken into account - even if only descriptively - in the balance sheet of benefits and costs.

In cases where an EIS or similar document is not required, the Determining Authority may attach certain conditions to the development consent or approval. For example, certain aspects of the development may need to be monitored carefully to ensure that adverse environmental impacts do not occur.

Directives for EIA

If the Determining Authority decides that a proposal is environmentally significant (including cases where the requirement is automatic), directives must be given for the form and conduct of the EIA. Scoping of the EIA should have been facilitated through the application of economic techniques.

The form and terms of reference for the EIA are drawn up by the Environment Authority in consultation with relevant government agencies and the proponent. The terms of reference should indicate the focus of the study, the time frame and the nature of recommendations.

The terms of reference may also stipulate requirements for economic analysis in the EIA. Some proposals do not lend themselves to economic analysis - for example, the construction of buildings in urban areas. In other cases it may not be worthwhile undertaking an economic analysis, as the benefits of resulting information may not match the investigational costs.

The proponent is responsible for hiring an appropriate project team. The required skills should be apparent from the initial screenings by the proponent and Determining Authority and from the guidelines issued by the Environment Authority.

Budgetary details can be determined by the proponent once the scope of the EIA, the time frame, research skills and necessary documentation have been established.

Conduct of the EIA - Economic Aspects

Tasks in the EIA

The EIA is carried out in an iterative manner, with close consultation between the proponent and Environment Authority. The initial investigations are directed towards the preparation of a draft EIS, which is then made available for public review. The final EIS is prepared after public submissions have been received, and after a final review by the Environment Authority.

The main tasks that should be carried out for the EIA are:

- identification of the physical and temporal boundaries of effect;
- investigation of proposed site(s) and surrounding physical, ecological, social and economic environment;
- conduct of baseline studies and identification of on-site and off-site environmental conditions, currently and in the future, without the proposed development;
- detailed analysis of the workings of the proposal, including its various phases, functions, and connections with the environment;
- investigation of environmental safeguards and mitigation measures;
- prediction of environmental impacts of the proposed development;
- analysis of alternatives to proposed action, including environmental effects;

- evaluation of development alternatives and environmental protection measures;
- indication of recommended course of action for implementation of the proposal, including environmental monitoring, environmental management and post-development provisions.

In dealing with the "economics" component of the EIA, it is assumed that the necessary physical, ecological and social investigations would have been carried out by appropriate specialists in the EIA team, and the results made available for economic interpretation. In carrying out economic sensitivity studies, economists in the team need to interact closely with the specialists in other disciplines.

It is beyond the scope of this book to discuss techniques for the design and on-site management of proposed developments, and the methods of analysing and predicting social and natural systems effects. Other texts and guidelines are available to support these tasks.

The main application of economic analysis in the EIA, as in the preliminary planning exercise, is to assess alternatives and support the design of environmental protection measures and management plans. Financial, fiscal and distributional effects may also be important considerations in evaluations of proposed developments, and are thus included.

Content of the Draft EIS

The content of the draft EIS will depend on the terms of reference for the EIA and on other legislative or administrative provisions. Typical requirements for content, for example as prescribed by Australian Commonwealth legislation, have already been listed. Such requirements may differ in other countries or jurisdictions.

The interaction of the Environment Authority and the proponent in preparing the draft EIS also depends on the practices and procedures laid down by government. The Environment Authority may consult with the proponent on the coverage of the draft EIS and provide guidelines for preparation. The Authority may consult with other individuals, experts or organisations in preparing these guidelines. It may also consult with the proponent to ensure that the draft EIS is suitable for public review.

Objectives of Proposal and Consideration of Alternatives

The specification of objectives for a proposal clearly affects the range of alternatives to be considered. The key questions are:

- should the proposal be undertaken or not?
- what is the optimal scale of the proposal?
- what is the optimal location?
- what is the best design and operating mode?
- what are the optimal environmental protection measures?

Objectives for development proposals from the private sector may be defined quite narrowly. If, for example, the objective is to produce petroleum products using a certain type of refinery at a particular site, the only alternatives available to the proponent may be to proceed with or abandon the project. Some flexibility will generally exist for the design of environmental protection measures. From a community viewpoint, broader alternatives for the proposed refinery could include other sites or imports of petroleum products. A private proponent can hardly be expected to examine alternatives that lie outside its interests or capacity. These must be considered by the Environment Authority or other government agencies.

For development proposals in the public sector, the range of alternatives may be wider. If the objective is to supply electricity to an urban area, the alternatives could include hydro or coal-fired power stations, alternative sites and the purchase of imported electricity.

Valuation of Benefits and Costs

Rough estimates of benefits and costs may have already been made in the plan formulation phase and the screening by the Determining Authority. In the EIA itself, more comprehensive and precise measurement of benefits and costs is required.

The key decision variable in economic evaluation is the present value of net benefits, as expressed in the formula in Chapter 1:

$$\mathrm{NPV} = B_d - C_d - C_p + B_e - C_e + B_o - C_o$$

Direct benefits and costs refer to the economic value of outputs and resources used in the proposal. Environmental protection costs are an additional cost item incurred within the proposed activity. Financial values generally provide a starting point for the estimation of economic values. Adjustments should be made, if necessary, for transfer payments, distortions of the exchange rate and wage rate, and capital constraints.

Environmental benefits and costs are "external" to the proposal. Although they are not appropriated or paid by the proponent, they do affect community economic welfare. Methods of valuing environmental benefits and costs are discussed in Chapter 3. It is suggested that the more easily applied techniques (market-based) be applied first. Surrogate market methods may be applied next. Contingent valuation may be used where necessary and where time and budget permit.

Valuation of cumulative or area-wide environmental impacts may be an important part of the valuation process. A major difficulty is predicting ultimate physical impacts within the regional environment.

Some kinds of cumulative effects - the direct and indirect impacts associated with regional production linkages - can be predicted through the use of generalised input-output models. These models yield estimates of pollutant discharges or other environmental variables for all activities in the region affected by the proposal. Natural systems models may have to be applied to predict the ultimate impacts. Once determined, the direct and indirect environmental impacts of the proposal can be valued in the usual way, using the economic valuation techniques described in Chapter 3.

It is important to measure environmental benefits and costs "with the proposal" in relation to the future conditions that are predicted to prevail "without the proposal", not in relation to the current set of environmental conditions.

If full valuation of all environmental effects is infeasible, it is desirable to fill in as much of the balance sheet as possible with economic measures, and otherwise include indicator variables or simple descriptions of effects. In extreme cases, it may be necessary to abandon economic valuation of environmental effects altogether and resort to cost-effectiveness analysis in the evaluation process (see Chapter 3).

Other external benefits and costs include costs of supporting infrastructure by government departments, such as roads and dams. Fiscal impact analysis is often a useful initial source of information to estimate benefits and costs in the public sector. Other external benefits and costs may also include income gains that accrue through the use of otherwise idle resources. The value of labour employed directly or indirectly by the proposal, that would otherwise remain unemployed in the surrounding region, constitutes a benefit of the project. Secondary economic impact analysis gives indications of such effects. The results of secondary impact calculations - for example, those derived from input-output analysis - must be interpreted carefully (see Chapter 4).

Future values of benefits and costs are converted to present values by applying a selected social discount rate. All values in the economic analysis should be in real prices rather than nominal prices. Inflation, that is, should be excluded from the analysis. The discount rate should reflect the social rate of time preference or the opportunity cost of capital. In general, the discount rate will be different from the market interest rate.

Evaluation of Alternatives

As explained in Chapter 1, benefit-cost analysis can facilitate the ranking of alternatives. The best alternative maximises the present value of net economic benefits.

Under certain conditions, alternatives should be ranked according to other criteria such as the marginal benefit-cost ratio. Standard texts in benefit-cost analysis explain when the use of these criteria is warranted.

If net benefits are negative for all alternatives, economic analysis suggests that the proposal should be abandoned. The proposal may, however, be accepted by government on grounds other than economic efficiency.

It may also be possible to modify the proposal in some way, to render it economically acceptable. Sensitivity analysis of key parameters of the proposal should help to indicate where changes in the proposal may lead to significant cost reductions or to increased benefits.

Environmental Protection Measures

The design of environmental protection measures involves a more restricted set of planning and management options within each general alternative for the development proposal. To achieve economic efficiency in the design of environmental protection measures, the net benefits should be maximised. In some instances, cost-effectiveness analysis, instead of full benefit-cost analysis, must be used.

If the Environment Authority insists on design or performance standards that are too costly, the proponent may abandon the proposal. Interaction at this stage of the EIA allows the proponent and Environment Authority to jointly determine an acceptable approach. Adoption of "best practicable means" technology for pollution control is one example of a mutually acceptable outcome.

The Environment Authority should be aware, in prescribing environmental protection measures for the proponent, that the objective should be the management of ambient environmental quality and associated damage costs. Directives such as maximum emission or effluent rates should be based on the predicted impacts on the environment. In the absence of specific information on environmental impacts, however, standard guidelines for emissions or effluents may be used.

Distributional Considerations and Incidence Analysis

Distributional aspects of the proposal may be indicated, especially the regional incidence of effects. Input-output analysis is one way of estimating the incidence of impacts on income and employment in the region affected.

Intergenerational Equity and Sustainable Development

In conventional benefit-cost analysis, use is made of a discount rate in evaluating planning and design alternatives. The effective life of a project in economic terms is strongly affected by the discount rate. It is most unusual to adopt an economic planning horizon greater than 50 or 60 years, even if the expected physical life of the proposed development, or the anticipated duration of environmental effects, is greater. For example, with a discount rate of 10%, a dollar's worth of benefits or costs in 50 years' time is worth only 0.85 cents in present value. The higher the discount rate used, the less important are future benefits and costs.

Conventional benefit-cost analysis would typically ignore long-term problems such as extinction of species, the future costs associated with the management of nuclear wastes, the delayed health effects of accumulated toxics in humans, and the risk of environmental disasters as structures approach the end of their useful life (for example, dams or chemical plants).

The proponent should not be obliged to choose the "correct" social rate of discount. If an official discount rate is prescribed, this should be used to derive present values. Sensitivity analysis (arbitrary selection of different rates of discount) will at least reveal whether changes in the social rate of time preference would lead to different rankings of alternatives and recommended actions.

The "new economics" of sustainable development recognises the importance of long-term benefits and costs, especially those associated with environmental effects. Distributional effects have always been acknowledged in conventional benefit-cost analysis. Sustainable development objectives focus attention on, among other things, the intergenerational distribution of benefits and costs. This matter has been discussed in Chapter 1.

One of the concepts of sustainable development is that natural capital (environmental assets) bequeathed to future generations should not be depleted or degraded. If future generations are predicted to suffer serious environmental losses or costs because the discount rate is "too high", the influence of the discount rate may be overruled and additional decision criteria introduced. The application of direct constraints or planning targets, rather than adjustment of the discount rate, is the preferred approach.

Such constraints may be applied for reasons other than intergenerational equity. The Environment Authority may insist on particular environmental protection constraints, regardless of the results of the economic analysis.

Fiscal Impact Analysis

Government revenues and outlays are an important element of the economic environment. Information on the fiscal impacts of a proposal may be crucial in deciding whether a proposal is acceptable. An analysis of fiscal impacts can also guide arrangements for transfer payments between the proponent and various government agencies and among government agencies themselves. Techniques of fiscal impact analysis are described in Chapter 4.

Fiscal impact analysis is sometimes confused with benefit-cost analysis in EIAs. The scope of fiscal impact analysis, however, is quite different; it concerns the financial costs and revenues of the public sector. For this reason it is sometimes called "cost-revenue analysis". Benefit-cost analysis deals with the economic significance of resource use from the perspective of society as a whole.

Fiscal impacts may be direct, as in the case of increased taxes, royalties or charges paid by the proponent to governments, or indirect - for example, increased outlays on water, sewerage or garbage services as a consequence of immigration to a locality induced by a major development. Local governments may have to respond to crisis conditions brought about by large-scale expansion in the surrounding region. Fiscal impacts may extend, however, to state or federal agencies.

To analyse fully the likely fiscal impacts of a proposal, the following tasks are required:

- projection of the costs and revenues of each governmental jurisdiction in the absence of the proposal;
- identification of the main categories of government revenue and expenditure directly related to the proposed development, and projections of the predicted transfer payments;
- identification of the indirect regional demographic and economic effects of the proposal, and the implications for costs and revenues in the different jurisdictions;
- aggregation of the direct and indirect fiscal impacts for each jurisdiction;
- description of any special funding arrangements.

If fiscal impacts are known in advance, governments will be able to respond in a more timely and effective manner. Local governments that offer inducements to industries to locate in the region should pay particular attention to the activities that may be encouraged indirectly. Sensitivity analysis can be carried out if there are uncertainties about the magnitude and timing of fiscal impacts.

The proponent may wish to offer compensation to government authorities if public finances are strained. Support by the proponent may be in-kind, such as the building of roads or provision of recreational amenities.

Dealing with Risk and Uncertainty

Risk and uncertainty may present problems in the analysis. As explained in Chapter 1, risk is measurable in economic terms when the probability distribution for an adverse event is known. Uncertainty prevails when the magnitude, mechanisms or probability of adverse events are not known.

The proposal may introduce new risks to the region - for example, as a characteristic of the production process (uranium mining, chemical plants, storage of natural gas) or by increasing the risk of natural disasters (stream modification that leads to increased flooding).

Risk assessment can assist in the evaluation of such risks. Much depends on social perceptions and attitudes towards risk. If society is considered to be risk neutral, expected values (arithmetic means) can be used for benefits and costs. Most people place a larger negative value on involuntary risk (for example, exposure to dam failure or to industrial accidents) than on voluntary risk. Even though it is difficult to measure the differences in the disutility of risk, the nature of the risk should indicate the direction of bias in the estimates.

There are several ways of handling the problem of uncertainty in a benefit-cost analysis:

- the economic analyst can conduct sensitivity analyses for effects that are uncertain;
- assuming that society is generally risk averse, "worst-case" values may be used for potentially adverse impacts;
- problems of uncertainty in the EIA can be pointed out in the EIS, allowing the Environment Authority (or the political system) to make a subjective judgement;
- decision analysis can be applied, thereby improving the quality of information contained in the EIS.

Irreversible Impacts

Avoidance of irreversible environmental damage supports the notion of sustainable development. Irreversibility is an economic concept which should show up in a benefit-cost analysis if it is properly conducted. Insistence on irreversibility effectively acts as a constraint within the analysis. It rests on the assumption that the goods and/or services provided by a particular set of environmental resources prior to a proposed development have sufficient social value to warrant their preservation or complete restoration at (or before) the end of the planning period. The benefits may not easily be measured.

If the cost of restoration or of some acceptable substitute is infinite, as would occur if a rare species is exterminated or a unique wilderness area destroyed, any proposed development resulting in such extinction or destruction could be unacceptable on economic grounds. The present value of net benefits would be (infinitely) negative, regardless of the discount rate.

Many adverse environmental impacts are reversible at a cost. The benefit-cost calculations in the EIS should reflect any existing legislation or regulations and any rulings by the Environment Authority. The Authority may insist on complete restoration at some future date (for example, rehabilitation of mining sites). The Environment Authority may require only partial restoration, or may permit a completely different subsequent use.

Possible tradeoffs of use and restoration of the affected environment during the course of the proposed development, and options for post-development use, can be assessed and presented in the EIS. Sustainable use can be made of the environment provided acceptable restoration is carried out.

Documentation of the Economic Analysis

The following guidelines should be observed in presenting the results of economic analysis in the draft EIS:

- all assumptions underlying the economic analysis should be stated;
- a presentation should be made of charts, diagrams, details of model simulations, base-case benefit-cost calculations and sensitivity studies;
- where a benefit-cost analysis has been conducted, impacts that could not easily be valued should be presented in descriptive form alongside the measured economic effects;
- an account should be given of the economic techniques used, together with sources of data;
- any reservations about limitations of the methods or accuracy of the results should be indicated.

Public Review of Draft EIS

Notice of Public Review

Except in rare cases where confidentiality of information in the EIS is judged necessary, the draft EIS is made available for public review. This may be done through a public consultation process including newspaper advertisements.

These public notices contain a brief description of the proposal, details of where the document can be purchased or read, an address for the lodgement of written submissions, and a closing date for receipt of submissions.

Purpose of Public Review

Public review is a mechanism of informing the public, in advance, of a proposed development. It allows the public to react to the proposal and to express community value judgements about the effects of the proposal.

Further information may also be obtained, especially from interested groups and parties with specialised knowledge of the local environment. Valuable submissions may be made by larger organisations, such as industry associations and conservation groups, with broad interests in environmental issues and skills in environmental assessment. Such organisations are becoming increasingly literate in the application of economics to environmental and resource management problems and in the provision of information on benefits and costs.

Handling of Public Submissions

Public submissions are examined by the Environment Authority, with referrals to other government agencies if required. The public submissions and comments by government are then passed on to the proponent.

Preparation of Final EIS and Completion of Decision Process

Submission of Final EIS to Environment Authority

All information and comments received after the public review must be considered by the proponent and incorporated in the final EIS. Public submissions may be reviewed jointly by the proponent and Environment Authority to modify the proposal if required. It is customary for public submissions to be published in the final EIS together with responses by the proponent. The final EIS is submitted to the Environment Authority for final review and recommendations.

Responsibilities of Environment Authority

In its review of the final EIS the Environment Authority is responsible for the following:

- ensuring that legislative requirements have been met and that all environmental matters have been properly dealt with;
- determining whether any additional environmental information is required, including data for monitoring before, during and after the development;
- formulation of recommendations or suggestions on the environmental aspects of the proposal, which may be applied in conjunction with approval of the proposal.

Submission to Determining Authority for Decision

If approved, the final EIS, together with all other information and responses to any other referrals from within government, is passed by the Environment Authority to the Determining Authority. Recommendations by the Environment Authority on the environmental aspects of the proposal are also conveyed. In some jurisdictions, communications between the Environment Authority and Determining Authority take place at Ministerial or higher government level.

Final Decision by Determining Authority

The Determining Authority makes a final decision and gives directives to proceed with, modify or (in rare circumstances) defer or cancel the proposal.

References

Ahmad, Y.J. and G.K. Sammy (1985) *Guidelines for Environmental Impact Assessment in Developing Countries*, Hodder and Stoughton, London.

Australian International Development Assistance Bureau (1989) *Environmental Assessment of Official Development Assistance: Environmental Screening Guidelines for Country Program Managers*, Canberra.

Basta, D.J. and B.T. Bower (eds) (1982) *Analyzing Natural Systems*, Resources for the Future, Washington D.C.

Biswas, A.K. and Q. Geping (eds) (1987) *Environmental Impact Assessment for Developing Countries*, Tycooly, Dublin.

Black, P. (1981) *Environmental Impact Analysis*, Praeger, New York.

Canter, L.W. (1977) *Environmental Impact Assessment*, McGraw-Hill, New York.

Clark, B.D., R. Bisset and P. Walthern (1980) *Environmental Impact Assessment: A Bibliography with Abstracts*, Mansell, London.

Erickson, P.A. (1979) *Environmental Impact Assessment: Principles and Applications*, Academic Press, New York.

FEARO (1985) *Environmental Assessment in Canada: Summary of Current Practice*, Canadian Council of Resource and Environment Ministers.

Jain, R.K., L.V. Urban and G.S. Stacey (1977) *Environmental Impact Analysis: A New Dimension in Decisionmaking*, Van Nostrand Reinhold, New York.

Leopold, L.B., F.E. Clarke, B.B. Hanshaw and J.R. Balsley (1971) *A Procedure for Evaluating Environmental Impact*, Circular No 645, U.S. Geological Survey, Washington. D.C.

Munn, R.E. (ed) (1979) *Environmental Impact Assessment: Principles and Procedures*, SCOPE Report No 5, Wiley, Chichester.

Nichols, R. and E.L. Hyman (1980) *A Review and Analysis of Fifteen Methodologies for Environmental Assessment*, Centre for Urban and Regional Studies, University of North Carolina, Chapel Hill.

O'Riordan, T. and R. Hey (1976) *Environmental Impact Assessment*, Saxon House, Farnborough.

Rau, J.G. and D.F.C. Wooten (1980) *Environmental Impact Analysis Handbook*, MacGraw-Hill, New York.

Chapter 3

ECONOMIC VALUATION TECHNIQUES

Considerable progress has been made in the development and application of economic techniques for the valuation of environmental impacts. The ease of measurement and the robustness of results depend to a large extent on the quality of data that are available. Market data form the basis of some of the techniques; in general these are the easiest to obtain. Where markets do not exist for environmental goods and services "surrogate market" values may be used. Where environmental values are entirely unpriced they may be derived by direct questioning using techniques known as "contingent valuation".

Frequently, it may not be possible, because of time and resource constraints or because of limitations in data and analytical methods, to obtain any economic measures of environmental values. Decisions can nevertheless be supported by the economic techniques of "cost-effectiveness analysis", "cost-tradeoff analysis" and "threshold analysis".

This chapter reviews the various economic valuation techniques, explains their theoretical underpinnings, and gives practical examples. Key references on environmental-economic valuation are provided in the main text. Other useful references include Cooper (1981), Dixon et al. (1986), Folmer and van Ierland (1989), Freeman (1979, 1982), Hyman (1981), James et al. (1978), Kerr and Sharp (1985), Kneese (1984), Krutilla and Fisher (1975), NOAA (1983), OECD (1974, 1976), Pearce (1978), Pearce and Markandya (1989), Sinden and Worrell (1979) and Yang et al. (1984).

D. James, The Application of Economic Techniques in Environmental Impact Assessment, 63–95.

The Productivity Changes Approach

Conceptual Basis

The productivity changes approach measures changes in the productivity of natural or man-made systems that result from changes in environmental conditions. Market prices can often be used to value these changes. The approach is useful for valuing environmental impacts that affect the productivity of fisheries, forests and agriculture or other capital assets.

Productivity changes may result from degradation of inputs to production or damage to outputs. Industrial production systems are often affected by a decline in the quality of environmental inputs. Highly saline intake water for an electric power plant, for example, may lead to accelerated corrosion of heat exchangers, poor system reliability and a reduction in electricity output.

The output of renewable production systems may be affected by a deterioration in environmental conditions. Vine crops damaged by fluorides are a relevant example. Other kinds of environmental degradation may result in a reduction in the value of fish catches, timber harvests and agricultural production. Human beings can also be viewed as a productive resource. Health damage from pollution may result in reduced labour productivity. This effect is described under the "loss of earnings" approach (see below).

Changes in net benefits usually consist of changes in producers' surplus. If affected producers are price takers, the change in producers' surplus is measured solely in terms of changes in output levels and cost conditions.

When a change in environmental quality results in a large change in output relative to the market for the good in question, the measurement of gross benefits must incorporate the change in consumers' surplus resulting from the price change. If the change in output also affects the market for factor inputs, measuring the effect on net benefits is a more complicated task.

An important requirement in the productivity changes approach is to identify carefully the "with and without" productivity effects. The term "with and without" does not mean "before and after". The magnitude of productivity changes is found by comparing the time-stream of output that would occur with and without the environmental effects of a proposed development. Projections or predictions must be made of future levels of output in both situations. Only the productivity changes directly attributable to the development activity are relevant.

Fisheries Example: A Loss of Output

A simple example of valuing productivity effects is shown in Figure 3.1. The output from a riverine fisheries industry is represented by X on the horizontal axis. At a certain level of water quality, the annual fish catch is OG. The price of fish is OP, so total returns to the industry equal the area OPBG. Average and marginal costs are OC. Total costs are thus represented by the area OCEG. Net benefits, consisting of producers' surplus, equal the area CPBE.

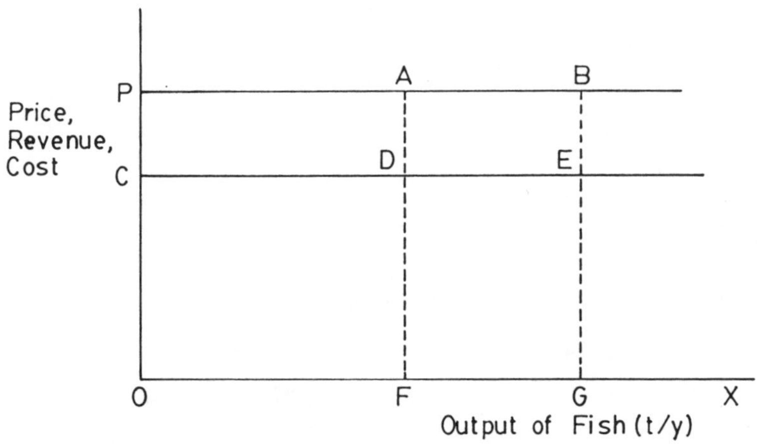

Figure 3.1. Economic effect of pollution on fisheries.

A proposal to establish a pulp and paper plant upstream is predicted to result in discharges of organic matter to the river, depleting oxygen levels and reducing fish stocks. This is expected to reduce fish catches to OF. Total revenue will equal the area OPAF, a reduction equivalent to the area FABG. There are now two possibilities:

1) Fishermen maintain the same fishing effort, such that total production costs are still represented by the area OCEG. The damage cost will consist of the change in net benefits, measured by the area FABG.
2) Fishermen reduce their inputs in proportion to the reduction in catch, in which case total production costs will equal the area OCDF. The damage cost is measured as the change in net benefits, or the area ADEB.

Conservation Farming Example: A Decline in Productivity

Soil erosion in upland areas is predicted to lead to declining crop productivity over time. In Figure 3.2 the predicted time trend in output is shown in AC. A change in farming practices, which involves no additional costs, is predicted to retard the decline in crop output, as shown by the line AB. The economic benefits of the improvement programme can be measured as the economic value of the savings in output, or the shaded area in the diagram.

Note that net benefits increase over time, even though the programme fails to reverse the downward trend in productivity. The total value of the change in farming practices can be measured as the present value of the change in net benefits up to a prescribed planning horizon such as H.

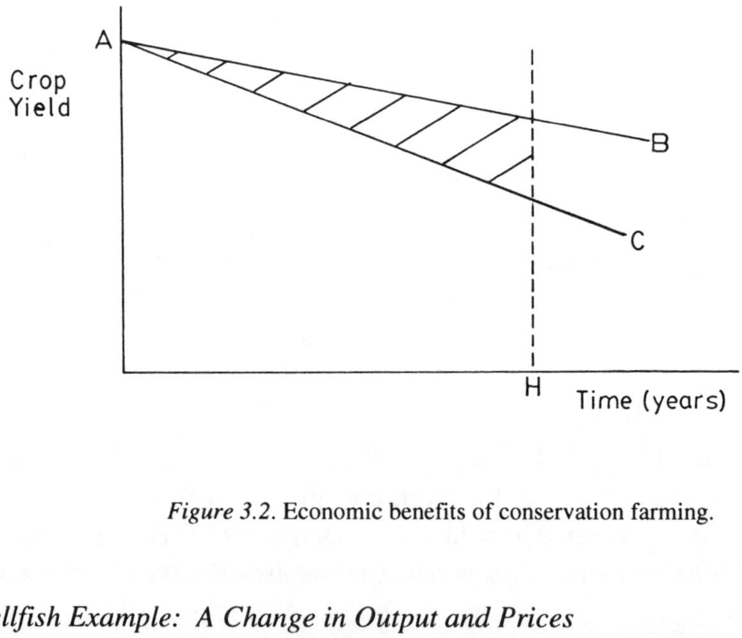

Figure 3.2. Economic benefits of conservation farming.

Shellfish Example: A Change in Output and Prices

A water pollution control programme in a coastal region is predicted to result in a significant increase in the productivity of the local shellfish industry. The long-run equilibrium market situation in the absence of the programme is described by the supply and demand curves S_1S_1 and D_1D_1 respectively, as shown in Figure 3.3. The equilibrium price of shellfish is OB, and the equilibrium quantity sold is OQ_1. Net benefits consist of the sum of consumers' surplus (area AFB) and producers' surplus (area BFD).

Following introduction of the water pollution control programme, productivity of the shellfish industry increases and the long-run supply curve shifts to S_2S_2. The equilibrium price falls to OC, and OQ_2 is the quantity sold. Net benefits now consist of consumers' surplus (area AGC) and producers' surplus (area CGE). Although there is an unambiguous increase in consumers' surplus, the change in producers' surplus depends on the position and slope of the industry supply curves with and without the pollution control programme.

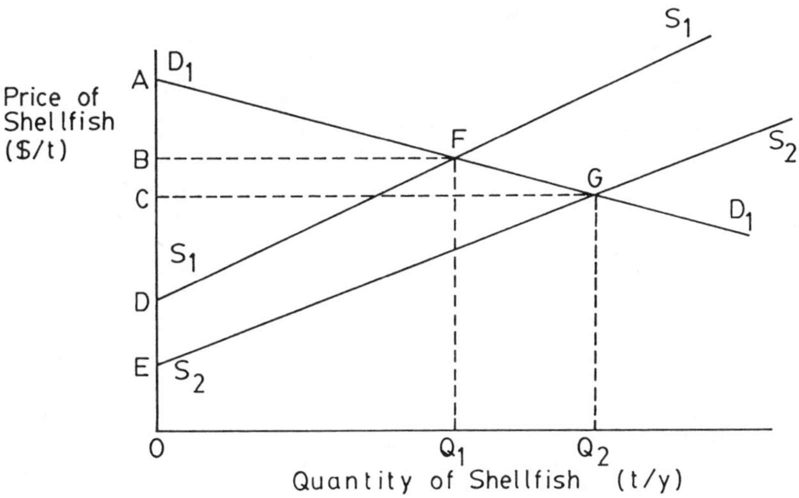

Figure 3.3. Pollution control and shellfish productivity.

Pollution Control Example: An Increase in Tourism Demand

The same water pollution control programme is predicted to lead to a significant increase in the demand for hotel accommodation, as more tourists are expected to be attracted to the region for recreational purposes.

In the absence of pollution control, the long run market situation for hotel accommodation is described by the supply and demand curves S_1S_1 and D_1D_1, as shown in Figure 3.4. Net benefits are measured by the area MTR, consisting of consumers' surplus (area MTP) and producers' surplus (area PTR).

With an improvement in water quality, the demand curve is predicted to shift to D_2D_2. Net benefits in the new situation would be measured by the area LVR, comprising consumers' surplus (area LVN) and producers' surplus (area NVR). The price paid per night and the number of visitor nights could both increase as a result of the pollution control programme.

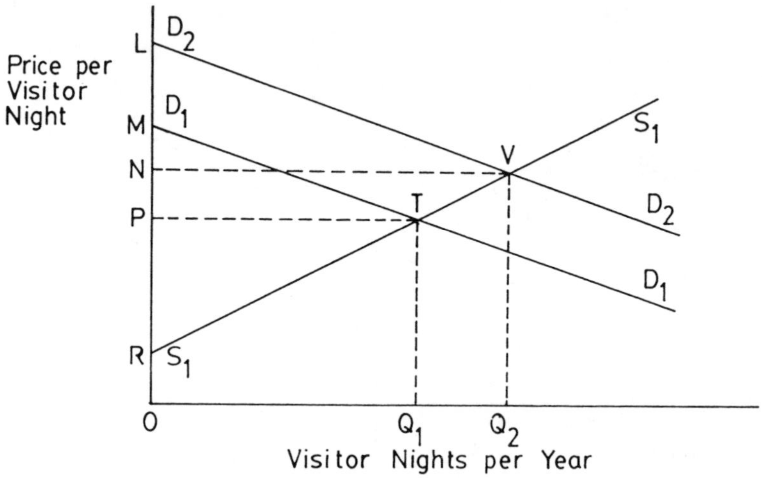

Figure 3.4. Pollution control and tourism demand.

Empirical Example: Rice Production

An example of the productivity approach can be found in a case study examining the cost effectiveness of various wastewater disposal alternatives for the Tongonan geothermal power plant in the Philippines (Grandstaff 1986). The productivity approach was used to value environmental damage costs associated with possible disposal of toxic waste water to rivers in the region.

In the study, it was assumed that the toxic elements in the discharged wastewater would induce farmers to abandon irrigation and turn to dryland cropping. Irrigated rice farming normally results in two crops per year. Dryland rice-growing gives only one crop and the yield per crop is much lower than for irrigated rice. Dryland rice production thus has much lower yields per hectare per year than irrigated rice.

The annual damage costs were measured as the reduction in net benefits resulting from the switch in production methods. It was assumed that market equlibrium prices would not change in response to the reduction in output, so existing prices could be used to value the productivity effects. The present value of the annual stream of damage costs over the life of the project (from years 3 to 30) was calculated using the official discount rate of 15%. The calculations are shown in Table 3.1.

Table 3.1.
Estimation of Rice Production Loss Resulting from Discharge of Toxic Wastewater to Mahiao River, Leyte (Pesos).

Total rice production area = 4,000 hectares
Net return/ha for irrigated rice = P 346
Net return/ha for non-irrigated rice = P 324
Annual loss by switching to dryland rice cropping = 4,000 x 346 x 2 - 4,000 x 324 = P 1.472 million
Present value of rice loss @ 15% discount for years 3-30 = 1.472 x 4.9405 = P 7.26 million

Source: Grandstaff (1986)

The Loss of Earnings Approach

Conceptual Basis

Air and water pollution can have serious effects on human health. One of the costs of ill health is the reduction in earnings associated with working days lost. Provided the relationship between adverse impacts of development and ill health can be established, it is a relatively straightforward task to estimate the value of lost income as a damage cost. Conversely, increases in earnings resulting from environmental mitigation measures or environmental improvement policies, programmes and projects, can be measured as economic benefits (Ostro 1983).

The loss of earnings approach measures only a minimum value of the costs of adverse health impacts. Effects on individuals too young to enter the work force will be excluded from the estimates, together with effects on those who are unpaid for their services such as housewives, and those in retirement. Medical expenses incurred as a result of ill health are also not included, although it is common practice to combine the loss of earnings with medical expenses to derive a cost of morbidity. Medical expenses can be viewed more realistically as a kind of replacement/repair cost for human beings adversely affected by environmental impacts. There is no allowance in the method for low-level impacts that do not lead to a loss of working days but nevertheless cause ill health and lower productivity. The psychic costs of pain and suffering are not accounted for in the estimates of lost income.

Some economists have attempted to use the loss of earnings approach as a means of valuing human life. This is clearly a spurious valuation procedure. It is generally agreed that an individual's willingness to pay to preserve his/her own life will be limited only by available income and assets, and the willingness to accept compensation to relinquish life will be virtually infinite. At best, attempts to value life should focus on the willingness to pay to reduce the risk of premature death. Statistical life, rather than the actual life of a particular individual, is the only meaningful concept in public decisions. Many economists consider attempts to value human life in monetary terms to be totally irrelevant, preferring to relegate the valuation problem to moral judgements rather than to economic assessment methods.

The loss of earnings is thus distinct from the value of life. The concept can be applied legitimately to changes in morbidity, but not to changes in mortality. Despite the partial nature of the results it yields, the loss of earnings approach can provide robust information on damage costs and environmental improvement benefits.

Method

To apply the loss of earnings approach, several steps are involved. They are illustrated here for adverse effects of pollution.

Beginning with the economic activity under consideration, the discharge of pollutants must be determined. Ambient concentrations are predicted from a dispersion model or reactive pollution model. The next step is to determine the population exposed to ambient concentrations, as well as the duration of exposure. Dose-response functions are used to predict the change in morbidity (and mortality) resulting from the change in levels of ambient concentrations. These effects are translated into the number of working days lost. In the last step, the loss of earnings is calculated using an estimate of the average wage rate or sickness compensation rate for the population affected, multiplied by the number of working days lost. More sophisticated approaches divide the population affected into specific socio-economic groups and apply the average wage for each group.

Table 3.2.
Morbidity Benefits from Reduction in SO4 Ambient Air Quality
(1974 meteorological year; $ million)

Country	Health Care Savings	Income Savings	Total Benefits
Austria	11 – 274	5 – 115	16 – 389
Belgium	23 – 573	12 – 311	35 – 883
Denmark	7 – 166	2 – 54	9 – 219
Finland	2 – 43	1 – 22	3 – 45
France	105 – 2,635	39 – 968	144 – 3,602
Germany	170 – 4,251	64 – 1,594	234 – 5,846
Netherlands	32 – 794	9 – 235	41 – 1,028
Norway	2 – 42	1 – 37	3 – 79
Sweden	6 – 156	3 – 76	9 – 232
Switzerland	14 – 355	6 – 142	20 – 497
United Kingdom	39 – 968	22 – 542	60 – 1,510

Source: OECD (1981)

Example: Sulphur Oxide Control

The steps described above were followed in a study by the OECD on the costs and benefits of sulphur oxide control in Europe, in which an attempt was made to value the health effects associated with sulphur oxide pollution (OECD 1981). The benefits of sulphur oxide control consisted of the damage costs avoided through a package of pollution control measures, described in the OECD report. Results of the study are presented in Table 3.2.

Using various types of dispersion models, the effect of each strategy on ambient concentrations throughout Europe was predicted. A linear dose-response relationship between sulphur oxide concentrations and mortality and morbidity effects was assumed. The analysts did not attempt to value mortality effects because of the controversy over methods for the valuation of human life.

The increase in earnings resulting from each pollution control strategy was calculated for each country according to the predicted reduction in morbidity and working time lost. Monetary valuations of these effects were based on national compensation of employees in each country on man-years lost for which sickness cash benefits were expected to be paid.

A reduction in health care costs was also included in the estimated benefits of each pollution control strategy. The total amount of health care savings was based on the total health care expenditure for each country, given as a percentage of GNP. The authors acknowledged that the estimate of total health costs based on mortality and morbidity would underestimate the true costs, as "psychic costs" would be ignored.

Practical Considerations

There should be little difficulty in placing economic values on morbidity effects, once these have been translated into working time lost (or saved). The main problem is obtaining appropriate dose-response date. This task is the responsibility of health experts, epidemiologists and medical researchers. Dose-response functions are typically derived from statistical studies estimating the relationship between exposure and morbidity/mortality. Difficulties may be caused by a lack of adequate data, the presence of multiple pollutants, confounding factors such as lifestyle or cigarette smoking, and the possibility of thresholds or nonlinearities in dose-response functions. Measuring the costs of chronic, non-debilitating diseases caused by pollution can be an additional problem.

Evidence from clinical tests and laboratory experiments can be used instead of statistical analysis to derive dose-response functions, although it is often difficult to project such results to large populations.

Valuing changes in mortality is a controversial topic. Although a loss of earnings will, in many cases, be a component of the value of human life, it is unlikely to be of much practical use in policy decisions. The number of lives saved or lost is best treated as a separate variable in assessing the effects of economic development and in designing environmental protection schemes and measures.

The Replacement/Repair Cost Approach

Conceptual Basis

The replacement/repair cost approach uses increased expenditure on the replacement, maintenance or repair of physical assets as a measure of environmental damage. The cost must be the result of some kind of environmental impact, as distinct from replacement and repair costs that would be normally sustained, given baseline environmental conditions.

The approach rests on the assumption that the replacement/repair cost is at least as high as the value of the asset that is damaged, otherwise it would clearly not be economic to contemplate replacement of the asset. The approach is applicable only where the magnitude of the damage can be measured and where the replacement cost can be estimated.

Example: Soil Conservation in Korea

An example of the replacement cost approach is documented in a case study by Kim and Dixon (1986) which examined the environmental quality aspects of upland agricultural projects in Korea (see Table 3.3). The soil in upland areas was being damaged by nutrient leaching and soil erosion. The cost of physically replacing the lost soil and nutrients was taken as a lower bound estimate of the value of measures designed to prevent the soil degradation.

Table 3.3.
Replacement Costs for Soil Erosion, South Korea

Nutrient	Kg lost per ha	Cost per Kg (won)	Cost per ha (won)
Nitrogen	15.7	480	7,536
Phosphorous	3.6	345	1,242
Potassium	14.6	105	1,533
Calcium	10.6	60	636
Magnesium	1.6	1,400	2,240
Organic Matter	75.4	175	13,195
Totals	121.5		26,382

Source: Kim and Dixon (1986)

Soil loss was measured in two sample survey areas by use of the "Universal Soil Loss Equation" which relates annual soil loss from a parcel of land to various explanatory variables. The general form of this equation is:

$$A = R * K * LS * C * P$$

where A is the annual soil loss in tons per hectare, R is a rainfall factor, K is a soil erodibility factor, LS is a length and slope factor, C is a cropping factor and P is an erosion control practice factor.

The annual costs of replacing the soil consisted of truck rental and labor costs for soil recovery and spreading. Nutrient losses were calculated using a lysimeter on trial plots. The annual replacement cost for nutrients was estimated using the market prices for fertilizers and the labor costs for spreading nutrients.

The Shadow Project Approach

Conceptual Basis

This approach is a special case of the replacement cost approach. It examines the cost of a hypothetical "shadow project" designed to provide substitutes for the environmental goods and services destroyed by a development activity. The costs of the shadow project can be added to the direct costs of the development activity. Development alternatives, including those which do not destroy the environment, can then be compared on the basis of net present value (NPV).

The shadow project approach requires two fundamental assumptions:

1) the value of the endangered environmental goods and services is greater than the costs of the shadow project;
2) the "shadow project" can adequately replicate the endangered environmental goods and services.

Example: Flood Protection in the Netherlands

The shadow project approach was used in the Netherlands to evaluate flood prevention schemes (Commissie Oosterschelde 1974). In the early 1950s, serious flooding occurred in the Oosterschelde estuary, causing extensive loss of property, crops, livestock and human lives. The Dutch government responded with an immediate plan to construct a dike which, although effectively preventing flood damage, would result in total destruction of the estuarine environment.

Public reaction to the plan induced the authorities to reconsider the estuary in terms of the important environmental services it provided such as recreation, commercial fisheries and habitat for birdlife. Alternative engineering solutions were examined. The major options comprised:

1) Complete closure of the estuary as in the original plan.
2) Complete closure of the estuary and construction of an artificial lagoon designed to replicate the environmental services of the estuary.
3) Strengthening of the existing system of dikes flanking the estuary.
4) Construction of a special dike with flood gates that would normally be left open, thus preserving the functions of the estuary, but could be closed during periods of storm and high tide, thus providing total safety for life and property.

It was not possible to place economic values on all benefits and costs of each alternative, especially environmental benefits and costs. Only those that had monetary or market values were included. Subjective judgements and tradeoffs were required to evaluate the options. Benefit-cost data for the various alternatives are shown in Table 3.4.

Table 3.4.
Benefit-Cost Comparison of Alternative Plans for
The Oosterschelde, Netherlands. Billions of 1974 Guilders. 4% Discount Rate.

	Complete Closure of Estuary	Complete Closure + Artificial Lagoon	Strengthen Existing Dikes	Partial Closure + Flood Control
Construction Costs	1.31	2.81	1.70	2.95
Operating Costs	0.02	0.02	0.08	0.02
Damage to Fisheries + Shellfish Production	0.47	0.47		
Cost of Water Management	0.10	0.10	0.06	0.06
Total Costs	1.90	3.40	1.84	3.03
Flood Prevention Benefits	1.92	1.92	1.37	1.89
Net Monetary Benefits	0.02	-1.48	-0.47	-1.14

Source: Commissie Oosterschelde (1974)

The original plan was subsequently rejected because the loss of environmental goods and services caused by blocking off the estuary was considered too great. The second plan was rejected because the net benefits were too low, and there was strong doubt that the lagoon could provide equivalent services of the estuary. The dike-strengthening option was rejected because of the high cost and high residual risk that the dike could still be breached, threatening life and property. The fourth option was eventually selected. Even though it was the most expensive, it gave complete flood protection and preserved the estuarine environment. The dike has since been constructed. It constitutes a major accomplishment of civil engineering.

The Relocation Cost Approach

Conceptual Basis

This approach examines the costs of relocating a physical facility which, because of a change in environmental conditions, can no longer operate effectively in its original location. These costs reflect the economic value of predicted environmental damage or, conversely, the benefits of preventing the damage.

Examples

A proposed development activity is predicted to discharge toxic effluent into a river. As a consequence, downstream irrigation or domestic water supply intakes might need to be relocated upstream of the discharge point. The costs associated with relocation of these facilities should be considered part of the damage costs of the proposed development activity.

A special case of the relocation cost approach concerns the forced relocation of people in order to accommodate a project. The construction of a dam could necessitate an "involuntary resettlement" of inhabitants. The costs of this project should include the costs of moving people and belongings to a new site, the costs of preparing a new site, legal and transfer costs. Although difficult to measure, psychic and other social costs should also be included, even if in a qualitative way.

The Defensive Expenditures Approach

Conceptual Basis

The defensive expenditures approach values the damage caused by environmental degradation according to the costs consumers and/or producers are willing to incur to prevent the damage. Individuals and producers will allocate resources to avoid adverse environmental impacts only when they consider the costs of preventing the damage to be less than or equal to the damage prevented. The willingness to incur costs to avoid damages resulting from environmental degradation is taken as an indication of willingness to pay for environmental protection. The summation of all defensive expenditures yields a minimum estimate of the value of the environmental damage. This information indicates the magnitude of the benefits of reducing the environmental damage.

In practice, undesired environmental impacts cannot always be avoided by averting behaviour on the part of individuals. The averting behaviour may not be a perfect substitute for environmental quality. The defensive expenditures approach may not capture the full decrease in utility resulting from a deterioration in environmental quality. For instance, household expenditures on noise exclusion facilities such as double glazing reduce only indoor noise. The disamenity of outdoor noise remains.

In many cases, several different types of averting behaviour may exist simultaneously. Averting behaviour can involve costs of time, money and other resources. An analysis using the defensive expenditures approach must identify all types of averting behaviour and the associated expenditure. In some instances, averting behaviour may yield secondary benefits other than those resulting from damage mitigation. In these cases, the defensive expenditures approach is less readily applied.

Example: Consumer Defensive Expenditures

A classic example of the defensive expenditures approach was the study of noise pollution in the vicinity of London's Heathrow Airport by Starkie and Johnson (1975). A linear regression model was used to link the demand for double-glazed windows with perceived noise levels in decibels. The model was based on the simple idea that households would be willing to buy "exclusion facilities" if the expenditure on these facilities was less than the value of the resulting reduction in noise. The demand for peace and quiet was estimated using the demand for noise exclusion facilities as a surrogate.

Example: Producer Defensive Expenditures

In certain instances environmental quality can be taken as a close substitute for an input in the production of a good or service. The savings in input costs can be taken as a lower bound estimate of the true benefits of improved environmental quality. For instance, municipal water supplies may require chlorination before use. An improvement in water quality may save chlorination costs per unit. The decrease in chlorination costs can be taken as a lower bound estimate of the improvement in water quality.

An example of the defensive expenditures approach can be found in the case study by Kim and Dixon (1986) of upland agriculture in Korea. It was discovered that lowland rice farmers were willing to incur the costs of constructing water diversion dikes as a defensive measure against siltation and crop damage caused by upland erosion. The defensive expenditures incurred were used as a minimum estimate of the damage costs inflicted on rice farmers by upland farmers.

The Travel Cost Method

Conceptual Basis

The travel cost method (TCM) was devised to value recreational sites, especially those of the "open access" variety. Users of the site pay nothing, or at most a nominal fee, thus there is no direct indication of the willingness to pay for user benefits.

TCM measures the willingness to pay for a site by deriving an imputed demand curve. Travel costs to the site are treated as a proxy for admission prices. The basic premise of TCM is that the costs incurred in travelling to a site have a direct influence on the number of visits undertaken. In other words, the number of visits to a site are a function of travel costs. Where the basic assumptions of the method are applicable, TCM is appropriate for the valuation of recreation sites, historic sites, and wilderness areas.

Method

TCM involves a number of steps, set out below:

1) Site visitors are surveyed to ascertain the frequency of visits and zones of origin. The visitation rate for each zone of origin is determined by dividing the number of visitors from each zone by the respective zonal population.

Population figures for each zone must be obtained from independent sources of data.

2) Travel costs to the site are determined for each zone. Travel costs should include all marginal costs of reaching the site, including the cost of travel time.
3) Visitation rates are regressed on travel costs across all zones to obtain the travel cost function. This function can be used to estimate visitation rates as a function of the "price" paid. Initially, the price paid by each zone will be the travel cost itself.
4) The assumption is then made that travel costs act as a proxy for admission charges to the site. An admission charge can be added to the travel cost for each zone and, using the travel cost equation, it is possible to "predict" the visitation rate for each zone.
5) For each simulated admission price, the predicted number of visits from each zone can be found by multiplying the population in each zone by the corresponding visitation rate. Total visits to the site, for the given admission price, can be determined by aggregating predicted visits across all zones. This gives one point on the implicit demand curve.
6) By repeating steps (4) and (5) the demand curve for site amenity can be constructed. The marginal willingness to pay (admission price) is given on the vertical axis and the number of visits on the horizontal axis.
7) Assuming a zero price is charged, total user benefits will consist of the area of consumers' surplus under the demand curve. The final figure represents the total willingness to pay for use of the site amenity. This value can be left as an annual benefit from the site, or it can be capitalised into a present value equivalent by dividing it by the appropriate discount rate.

More complex travel cost models exist which take into account the availability of substitute sites, differences in tastes and incomes across zones, the fact that visits may be part of a multiple destination trip, and the problem of site congestion.

Example

A good example of TCM is the study of Lumpinee Park in Bangkok, conducted by Grandstaff and Dixon (1986). Following the method described above, a survey of visitor use was first undertaken. Population and visitation rates for the various zones are given in Table 3.5. Travel costs were estimated for each zone, based on the distance travelled and the time taken. The best-fitting equation for site visitation rates was found to be:

$$V = 1322.68 - 58.464\ TC$$

where V is the visitation rate per 1000 population per year and TC is average total travel cost.

For successively higher prices, and making use of the regression equation, visitation rates and the corresponding number of visits were estimated, as shown in Table 3.6. This information was then used to derive the demand curve (see Figure 3.5). The area under the demand curve was found by simple trapezoidal integration. The study indicated that the annual user benefits would be 13,204,588 Baht.

Table 3.5.
Visitation Rate per 1,000 Population per Year.
All Zones: Lumpinee Park, Thailand.

		Sample		
Zone	Population	Persons	%	Visitation Rate/1000
1	190,450	10	5.3	590
2	235,647	27	14.4	1,288
3	77,112	4	2.1	583
4	131,542	23	12.3	1,965
5	380,416	30	16.0	886
6	519,869	18	9.6	389
7	523,831	37	19.8	794
8	123,109	3	1.6	274
9	479,659	4	2.1	94
10	201,334	4	2.1	223
11	388,333	6	3.2	174
12	255,555	7	3.7	308
13	262,097	1	0.5	43
14	140,249	4	2.1	320
15	382,621	5	2.7	147
16	204,434	1	0.5	55
17	113,769	3	1.6	296
Total	4,610,027	187	99.6[a]	

[a] Precentages do not add up to 100 because of rounding

Source: Grandstaff and Dixon (1986)

Table 3.6.

Annual Visits to Lumpinee Park at Various Admission Fees

Zone	Cost per Visit	Number of Visits at Various Admission Fees (Baht)					
	(Baht)	0	4	8	12	16	20
1	8.08	161,978	117,438	72,900	28,362		
2	3.72	260,483	205,375	150,268	95,160	40,052	
3	10.25	55,800	37,768	19,734	1,700		
4	5.04	135,254	104,492	73,730	42,968	12,206	
5	8.64	311,085	222,123	133,160	44,197		
6	10.00	383,788	262,213	140,639	19,064		
7	13.66	274,624	152,123	29,621			
8	16.65	43,021	14,231				
9	14.18	236,884	124,713	12,541			
10	15.50	83,893	36,810				
11	20.35	51,701					
12	19.52	46,424					
13	17.16	83,776	22,482				
14	17.01	46,058	13,260				
15	18.43	93,890	4,412				
16	27.59						
17	21.70	6,167					
T. Visits/Yr		2,274,826	1,317,440	632,593	231,451	52,258	0

Source: Grandstaff and Dixon (1986)

Methodological Issues

Estimates derived from TCM are site specific and measure only the consumer surplus of present users. So-called "non-user" benefits such as existence value and option value are not captured. Furthermore, it is possible that the relative price of recreational sites may increase in the future. It is therefore reasonable to assume that the travel cost method will yield only a minimum value for site benefits. Allowance can be made for an increase in the number of future users and the estimated consumers' surplus, based on population projections and the

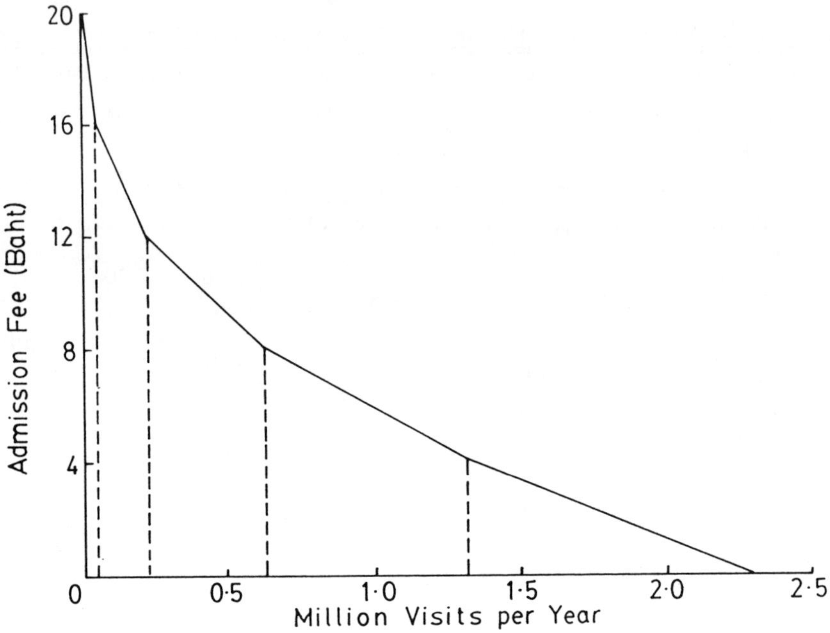

Figure 3.5. Demand for recreation: Lumpinee Park. (Source: Grandstaff and Dixon (1986)).

income elasticity of demand for recreation dependent on the natural environment.

Although conceptually straightforward, TCM may present some statistical problems, particularly in the proper specification of the functional form of the travel cost equation. Conducting the site survey and collecting and processing the necessary data could be expensive and time-consuming.

The Property Value Approach

Conceptual Basis

The property value approach attempts to measure environmental damage costs or environmental improvement benefits by tracing the effect of environmental quality on property prices. It is based on the economic concept that the value of a property is directly related to the present value of the expected stream of benefits to be derived from that property. The method has been applied extensively to residential housing, but can also be applied to land and other types of property. An essential requirement for the method is that people use the property and are therefore affected, either favourably or unfavourably, by the surrounding environment.

The price that people are prepared to pay for property depends on the package of attributes that each property contains, such as its size, the number of rooms and construction materials. One such attribute is environmental quality. It follows that, all other factors remaining constant, an improvement in environmental quality should lead to an increase in property values for the properties affected. The increase will represent the present value of the higher "economic rent" that people can expect to enjoy over the life of the asset. In theory, the willingness to pay this additional capitalised value should yield a measure of the total benefits from the improvement in environmental quality. The marginal willingness to pay consists of an implicit or "hedonic" price for environmental quality (Freeman 1979).

Method

The property value method is usually applied in two stages. First, the prices of houses are found as a function of attributes using regression analysis. This is based on an empirical equation of the form:

$$P = f(A_1, A_2, \dots, A_n, Q)$$

where P is the property price, f is the property value function, $A_1 \dots A_n$ are property attributes and Q is environmental quality. The property price function for environmental quality indicates the effect of Q on P, as shown in Figure 3.6.

In the second stage, an attempt is made to determine the change in community benefits that will result from a change in environmental quality. The slope (partial derivative) of the property price function with respect to environmental quality yields a curve reflecting the marginal willingness to pay for environmental quality within the property market (see Figure 3.7).

The shape of the marginal willingness-to-pay curve depends on the type of mathematical function that is specified in the regression analysis. To obtain a downward-sloping marginal willingness-to-pay curve, such as that shown in Figure 3.7, a nonlinear regression equation must be imposed on the data. If the regression equation is linear (uses fixed coefficients) the marginal willingness

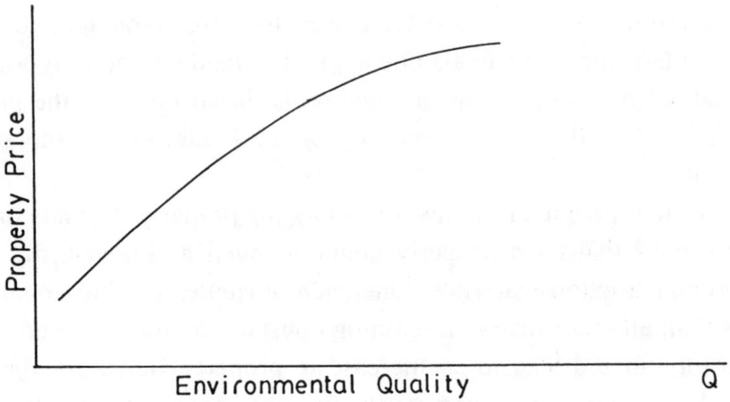

Figure 3.6. Property prices and environmental quality.

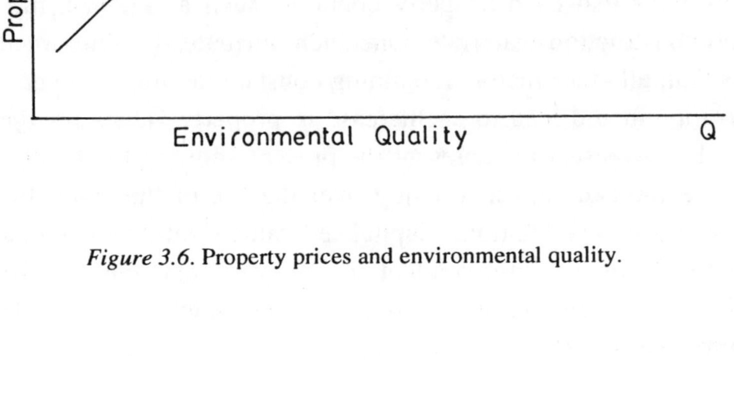

Figure 3.7. Household demand and marginal willingness to pay for environmental quality.

to pay curve will be a horizontal line.

The price that each household is prepared to pay for each level of environmental quality is indicated by an "inverse demand curve". A typical relationship between household inverse demand curves and the rent function is shown in Figure 3.7. To maximise utility, each household will purchase environmental quality up to the point where the marginal willingness to pay equals the market price. Changes in household benefits are measured by the areas under the household inverse demand curves, over the range of the change in environmental quality. To derive an estimate of the change in community benefits, these areas should be summed over all households affected.

The difficulty at this point is that there is still no information on the true household demand curves for environmental quality. The marginal willingness-to-pay curve derived from market data reflects the equilibrium prices that have occurred for all sales in the property market. The analyst does not know how the supply and demand for properties have responded to market prices, or the nature of household demand curves for environmental quality itself. In econometrics, this is known as the "identification problem".

To resolve this difficulty, it is necessary to introduce arbitrary assumptions or more powerful regression techniques (Hufschmidt et al. 1983). The possibilities include the following:

1) Assume a constant marginal willingness-to-pay function for environmental quality; this implies that a linear regression equation has been used for the property price function.
2) Assume that all households have identical utility functions and incomes. In this case, the marginal willingness-to-pay curve becomes the inverse demand curve for each household.
3) Apply regression analysis to the data for individual households assuming (a) the supply curves for properties with specific bundles of attributes are horizontal or (b) the supply curves are vertical. The slopes of the household demand curves for environmental quality can be estimated from this analysis.
4) Apply regression analysis to the household data, using simultaneous equation methods. This will identify the separate supply and demand functions.

In practice, the approach most commonly used is (3b). The observed marginal willingness to pay for environmental quality by each household is regressed on relevant household variables such as the level of environmental quality experienced by the household and income earned. The resulting function yields a representative inverse demand curve for environmental quality for all households in the sample.

The values generated by the property value technique are capitalised values. These capitalised prices can be converted into annual benefits by applying a social discount rate.

Choosing locations with less environmental degradation is only one way to mitigate the effects of a deterioration in environmental quality. A range of "averting behaviour" may exist as an alternative, such as installation of filters and double glazing. If such expenditures lead to an increase in property values, they should be classified as costs of environmental deterioration and added to the estimates obtained from a property price study. If such expenditures do not affect property values they can be ignored.

Data, time and computer requirements are substantial for this technique. Furthermore, the statistical problems of correctly identifying and estimating the hedonic price function can be severe. Some analysts have attempted to preserve the central concept of the property price technique, but instead of relying on statistical analysis, have questioned property agents, who usually possess sound information on what people are prepared to pay for various types of property attributes.

Contingent Valuation Methods

Conceptual Basis and Context

Contingent valuation methods have the advantage of being applicable in situations where no market data are available. For many kinds of environmental values they are the only techniques that can be used (Cummings et al. 1986).

The contingent valuation approach rests on the assumption that consumers can and will reveal their true willingness to pay for nonmarket goods, contingent upon a hypothetical market (Mitchell and Carson 1989; Wilks 1990). The method assumes that persons respond to the contingent situation just as in a real market, where the usual conditions of consumer behaviour hold, such as utility maximisation subject to a budget constraint.

Contingent valuation methods are commanding increasing attention as a means of valuing environmental effects. They are being used increasingly to value natural areas (Lockwood and De Lacy 1992). They have been applied by the Australian Resource Assessment Commission in major inquiries on Kakadu Conservation Zone and Australian forests (RAC 1991, 1992).

One important area of application is in the assessment of "option values" and "existence values" (Bishop 1982). An option value refers to the price that people are willing to pay to maintain the possibility of using a resource such as a wilderness area. It may be likened to an insurance premium. Existence values refer to the satisfaction people receive from the knowledge that a natural resource (for example, whales or conservation areas) will be preserved. Option values and existence values are usually not expressed through market mechanisms. Contingent valuation methods offer some hope of actually measuring these values.

A range of alternative methods exists for eliciting contingent bids. These methods measure either consumer willingness-to-pay for a good or service or consumer willingness-to-accept compensation for loss of that good or service. The diversity of methods is useful in that it allows various assumptions to be tested. A range of such methods is described below.

Bidding Games

The most common set of CVM approaches is referred to as "bidding games". The simplest is the open bid. Subjects are asked how much they are prepared to pay to enjoy a particular environmental amenity or bring about some improvement in the quality of the environment. Convergent bidding is a different version of bidding games. Each subject is given a high value which exceeds his/her maximum willingness to pay and a low value which almost certainly will be paid. The higher values are reduced and the lower values are increased until they converge on an equilibrium value.

Under another CVM technique known as "dichotomous choice" subjects are asked whether they would pay a certain price or not. A different price is suggested to each subsample that is drawn from the population. A cumulative frequency distribution is then fitted to the responses, from which the mean and median values for willingness to pay can be calculated.

Trade-Off Games

Another form of the contingent valuation method is the trade-off game. Each subject is asked to rank combinations of two goods, such as a sum of money in the form of a gift and some environmental attribute (air of a certain quality or the number of visits to a natural area). For any pair of combinations the subject must indicate whether he/she prefers one over the other or whether he/she is indifferent between them. The marginal rate of tradeoff of money for environmental amenity can be assessed at the point of indifference.

If the initial trial does not result in such a point, the exercise is repeated using different combinations of money and the environmental attribute. The process continues until the individual's point of indifference and hence marginal value of the environmental attribute is reached. Tradeoff games can be extended by including more than one environmental attribute. The aim is to find the combinations to which the individual is indifferent and determine the marginal valuations of environmental attributes in each game.

Costless-Choice

The costless-choice method is similar to the tradeoff games approach. Instead of using money as one of the goods, only marketed commodities (such as a bicycle, radio or some other consumer good) are used in combination with environmental attributes. At the point of indifference, the individual's marginal rates of tradeoff among all goods are ascertained. The monetary valuation of environmental attributes is obtained by substituting the market value of the conventional goods chosen. The method relies on the assumption that the individual knows the true market values of the conventional goods.

Priority Evaluator Method

The priority-evaluator method simulates the standard microeconomic model of consumer behaviour. Sinden and Worrell (1979) have explained the method, using a simple example. A sample of subjects is first drawn. Each individual is given a hypothetical sum of money to spend on conventional goods and environmental attributes at assumed prices. The game might, for example, contain five objects (four conventional goods and one environmental attribute) in three different quantities, giving fifteen possible choices. The budget must be large enough to permit each individual to buy positive amounts of all goods and the environmental attibute, but not large enough to buy the maximum amounts on offer in the game. A constrained choice must therefore be made.

The game seeks a typical individual in the market. For each individual to maximise his/her utility, the marginal utilities of each quantity purchased should be proportional to the prices. If the prices set in the game were those maximising utility, each of the fifteen possible choices would have an equal chance of being selected. The ratio of the observed frequency of choice to the expected frequency would be 1 for each of the fifteen possibilities. If the ratio exceeds 1 for any possible choice, the initial price was clearly set too low. It must be adjusted upwards. If the ratio is less than 1, the initial price was too high and should be adjusted downwards. By changing the prices and repeating the trial until ratios close to 1 are obtained for all fifteen possibilities, the true preferences and marginal valuations of the typical individual can be derived.

Sources of Bias and Other Considerations

The aim of CVM is to simulate a market for goods that are not marketed. An important requirement for CVM is that this hypothetical market resembles a real market as closely as possible. However, various sources of bias may exist that can influence consumer bids (Bishop and Heberlein 1979; Mitchell and Carson 1989).

For instance, there may be an incentive for respondents to overstate or understate their willingness to pay or willingness to accept compensation. This phenomenon is known as "strategic bias". If there is reason to believe that no actual payment will be elicited, respondents may overstate their willingness to pay. Conversely, they may understate their valuation because they have reason to suspect that a user charge may actually be imposed. There is little evidence of strategic bias either in laboratory experiments or in actual CVM studies (Pearce and Markandya 1989).

The quality, quantity and presentation of information in a CVM study can also influence bids. Changes in resource availability or environmental quality must be conveyed to the respondent in a manner that is both understandable and technically precise. The use of visual aids can prove useful in this respect.

Starting point bias could be a problem in bidding games. This occurs where the suggested value in the initial bid may focus the responses around that level.

Another kind of bias is compliance bias, which refers to the subject's desire to please the questioner. The willingness to pay for environmental amenities, for example, may be overstated because the subject considers it a "good thing" to nominate a high value.

An important source of bias stems from the choice of payment vehicle or instrument. The kind of charge or compensation payment - for example, general taxes, a gate charge, a subsidy to polluters, or lump sum payments to individuals - can influence the magnitude of bids. The hypothetical payment vehicle chosen should enable respondents to express realistic values.

Mitchell and Carson (1989) argue that two criteria should be met in choosing a payment vehicle: realism; and neutrality. In particular, the payment vehicle should not elicit an unfavourable response. Taxes often fail to pass the second criterion; people can react emotively to any suggestion of new or increased taxes.

Various problems with contingent valuation methods have yet to be resolved. For example, economic theory suggests that an individual's willingness to pay for the use of an environmental amenity should be approximately the same as his/her willingness to accept compensation to forgo that use. In theory, the willingness to accept compensation might be slightly greater as a result of the so-called "income effect". In certain experiments it has been found that individuals place a significantly higher value on losses than on equivalent gains (Knetsch 1990). This finding appears to undermine a fundamental assumption in benefit-cost analysis that benefits are costs avoided and costs are benefits forgone. The full implications of this research have yet to be determined.

Cost-Effectiveness Analysis

In many circumstances it may be difficult to estimate the benefits of environmental protection. A useful economic approach under such circumstances is cost-effectiveness analysis. The method involves setting a goal in the form of a physical target, social indicator, health standard, or environmental quality standard, then finding the least-cost way of achieving it. The benefits are not specified in economic terms.

Environmental quality targets are typically established in the form of standards or tolerance levels for humans, plants, animals and property. If environmental goals cannot be met, the planned development should be cancelled. Alternatively, a set of remedial measures may be considered that permit environmental standards to be reached. This usually adds to the cost of development. In some cases, the additional cost may prohibit adoption of the proposal.

Cost-effectiveness analysis requires identification of all possible ways of meeting environmental quality targets while producing a prescribed set of economic goods and services. The cost of each alternative is assessed (direct production costs and environmental protection costs), leading to selection of the set of measures that involves the overall least cost. The search for the least-cost solution may be conducted manually or with the assistance of computerised mathematical models. The approach is especially useful in situations where economic activities are already established, and the aim of policy is to protect or enhance the environment while maintaining existing lines of production.

Cost Tradeoff Analysis

When environmental quality targets cannot easily be established, subjective judgements about the optimal level of environmental protection can be assisted by cost tradeoff analysis. In the case of pollution control, the shape of the abatement cost curve may suggest a logical cut-off point. Figure 3.8 gives an example, for the abatement of emissions from an industrial plant. Reducing emissions by OM involves only a modest total economic cost. Similarly, total costs increase only moderately when a level of ON is achieved. However, total costs rise sharply when a level of OP is reached. The marginal cost of abatement jumps to a high level between N and P. The additional benefits of control would need to be very large to justify an abatement target of OP. Because the benefits are unknown, an arbitrary or subjective guess must be made about the marginal benefits.

In general, the marginal benefits will be smaller when there is almost no remaining pollution, than when the first steps are taken to improve a bad environmental situation. Unless there were evidence to indicate otherwise, an abatement level of ON would seem to be an optimal level to choose. The resources that might have been employed in moving from ON to OP could be used to produce other economic goods and services, or to protect the environment somewhere else.

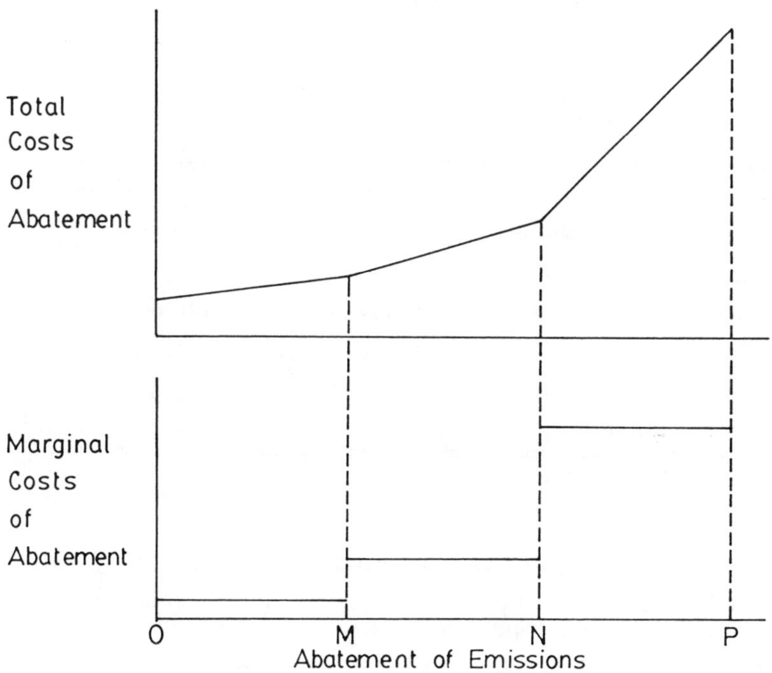

Figure 3.8. Cost tradeoffs for emission abatement.

The above discussion has shown how abatement costs can be traded off against some kind of environmental indicator - in this case, emissions discharged from an industrial plant. Other indicators might take the form of ambient concentrations of residuals, or levels of receptor response, such as the probability of death or disease in the population. Decisionmakers frequently place implicit values on human life and health when they stipulate cut-off points for public expenditures on safety precautions, health programmes and medical facilities. Tradeoffs may involve a single indicator or a collection of indicators, as in the case of regional environmental management where many types of pollutants may need to be considered. Simulation models can be used to study the relevant tradeoffs.

A practical example of cost tradeoffs with a single indicator is given in Table 3.7 for the control of sulphur oxides from electric power stations (Ruben 1983). The figures apply to a typical 2,000 MW station in Japan, but the costs have been converted to 1982 $US. The plant uses coal with 1.2% sulphur and 15-20% ash, costing $70 per tonne. The plant capacity factor is 70%. The flue gas desulphurisation scrubbers, which remove SO_x, have an estimated life of seven years. A discount rate of 10% is used.

Table 3.7.

Capital Cost ($/Kw) for SO_x Controls for a New Coal-Fired Power Plant in Japan

SO_x Emission (ppm)	SO_2 Removal Rate (%)	Capital Cost	Marginal Capital Cost ($/ppm)
1,000	0	0	
200	80	134	0.16
100	90	154	0.20
50	95	182	0.56

Source: Derived from Ruben (1983)

Ruben's research showed that SO_x removal in general is very expensive, costing from 15% to 18% of total plant cost. The results in Table 3.7 are derived from data published by Ruben. The marginal cost of achieving an 80% rate is $0.16 per ppm reduction (134/800); for a 90% rate it is $0.20 (20/100); and for a 95% rate it rises dramatically to $0.56 (28/50). These calculations suggest that, once a scrubber is planned, it is probably worthwhile aiming for a 90% removal rate. A higher removal rate would probably be prohibitively expensive.

Threshold Analysis

Conceptual Approach

In cases where development and environmental preservation - for example of unique forests or wilderness areas - are mutually exclusive alternatives, it may be difficult to obtain reliable estimates of projected community values for the natural environment. The decision model that can be used under these circumstances relies on the concept of "threshold values". The approach was developed by Krutilla and Fisher (1975) for a proposed hydro power scheme in a wilderness area known as Hell's Canyon.

To apply the threshold value model, the first step entails estimating the NPV of the development option - such as the economic benefits of logging an old-growth forest. The next step builds on the observation that as the population expands and real incomes increase, people reveal a proportionally greater preference for the use or preservation of natural environments. This trend is evident in expenditure patterns for outdoor recreation, membership of conservation groups, ownership of recreation vehicles and equipment, and purchases of environmental books, films and other materials. This step results in an estimate of the projected rate of growth of environmental benefits, even though the absolute level may not be known.

The final step involves a comparison of the NPV of the development option and the preservation option, to calculate the capital sum (the threshold value) required to make the preservation option as economically attractive as the development option. With an estimated growth rate for future environmental benefits, it is also possible to calculate the intial year's payment needed to switch from the development option to the preservation option. This value can be calculated as a total payment or as a per capita payment for the relevant population.

The information derived from the threshold value model does not provide an explicit value for the environment, but it does indicate to policymakers the minimum price the community could be expected to pay to justify preservation of a natural environment. The information is presented to decisionmakers in a form that may assist them to make a subjective judgement about the implicit value of the environment to the general community.

References

Bishop, R.C. (1982) 'Option Value: An Exposition and Extension', *Land Economics*, 58:1-15.

Bishop, R. and T.A. Heberlein (1979) 'Measuring Values of Extramarket Goods: Are Indirect Measures Biassed?' *American Journal of Agricultural Economics*, 60:962-30.

Cooper, C. (1981) *Economic Evaluation of the Environment*, Hodder and Stoughton, London.

Commissie Oosterschelde (1974) *Rapport*, Staatsuitgeverij, 's Gravenhage.

Cummings, R.G., D.S. Brookshire and W.D. Shulze (1986) *Valuing Environmental Goods: An Assessment of the Contingent Valuation Method*, Rowman and Allenheld, New Jersey.

Dixon, J.A., R.A. Carpenter, L.A. Fallon, P.B. Sherman and S. Manopimoke (1986) *Economic Analysis of the Environmental Impacts of Development Projects*, Earthscan, London.

Dixon, J.A. and M.M. Hufschmidt (1986) *Economic Valuation Techniques for the Environment: A Case Study Workbook*, Johns Hopkins University Press, Baltimore.

Folmer, H. and E. van Ierland (eds) (1989) *Valuation Methods and Policymaking in Environmental Economics*, Elsevier, Amsterdam.

Freeman, A.M. (1979) *The Benefits of Environmental Improvement*, Johns Hopkins University Press, Baltimore.

Freeman, A.M. (1982) *Air and Water Pollution Control*, Wiley Interscience, New York.

Grandstaff, S. (1986) 'Tongonan Geothermal Power Plant in Leyte, Philippines', in J.A. Dixon and M.M. Hufschmidt (eds) (1986) *Economic Valuation Techniques for the Environment: A Case Study Workbook*, Johns Hopkins University Press, Baltimore.

Grandstaff, S. and J.A. Dixon (1986) 'Evaluation of Lumpinee Public Park in Bangkok, Thailand, in J.A. Dixon and M.M. Hufschmidt (eds) (1986) *Economic Valuation Techniques for the Environment: A Case Study Workbook*, Johns Hopkins University Press, Baltimore.

Hufschmidt, M.M., D.E. James, A.D. Meister, B.T. Bower and J.A. Dixon (1983) *Environment, Natural Systems and Development: An Economic Valuation Guide*, Johns Hopkins University Press, Baltimore.

Hyman, E.L. (1981) 'The Valuation of Extramarket Benefits and Costs in Environmental Impact Assessment', *Environmental Impact Assessment Review*, 2:226-258.

James, D.E., H.M.A. Jansen and J.B. Opschoor (1978) *Economic Approaches to Environmental Problems*, Elsevier Scientific, Amsterdam.

Kerr, G.N. and B.M.H. Sharp (1985) *Valuing the Environment: Economic Theory and Applications*, University of Canterbury, Christchurch.

Kim, S.H. and J.A. Dixon (1986) 'Economic Valuation of Environmental Quality Aspects of Upland Agricultural Projects in Korea', in J.A. Dixon and M.M. Hufschmidt (eds) (1986) *Economic Valuation Techniques for the Environment: A Case Study Workbook*, Johns Hopkins University Press, Baltimore.

Kneese, A.V. (1984) *Measuring the Benefits of Clean Air and Water*, Resources for the Future, Washington D.C.

Knetsch, J. (1990) 'Environmental Policy Implications of Disparities between Willingness to Pay and Compensation Damaged Measures of Values', *Journal of Environmental Economics and Management*, 18:227-237.

Krutilla, J.V. and A.C. Fisher (1975) *The Economics of Natural Environments*, Johns Hopkins University Press, Baltimore.

Lockwood,M. and T. De Lacy (1992) (eds) *Valuing Natural Areas: Applications and Problems of the Contingent Valuation Method*, Johnstone Centre of Parks, Recreation and Heritage, Charles Sturt University, Albury.

Mitchell, R.C. and R.T. Carson (1989) *Using Surveys to Value Public Goods: The Contingent Valuation Method*, Resources for the Future, Washington D.C.

NOAA (1983) *Assessing the Social Costs of Oil Spills*, National Oceanic and Atmospheric Administration, Rockland Md.

OECD (1974) *Environmental Damage Costs*, Organisation for Economic Cooperation and Development, Paris.

OECD (1976) *Economic Measurement of Environmental Damage*, Organisation for Economic Coperation and Development, Paris.

OECD (1981) *The Costs and Benefits of Sulphur Oxide Control*, Organisation for Economic Cooperation and Development, Paris.

Ostro, B. (1983) 'The Effects of Air Pollution on Work Loss and Morbidity', *Journal of Environmental Economics and Management*, 10:371-382.

Pearce, D.W. (1978) *The Valuation of Social Cost*, Macmillan, London.

Pearce, D.W. and A. Markandya (1989) *Environmental Policy Benefits: Monetary Valuation, Organisation for Economic Cooperation and Development*, Paris.

Resource Assessment Commission (1991) *Kakadu Conservation Zone Inquiry: Final Report*, Australian Government Publishing Service, Canberra.

Resource Assessment Commission (1992) *Forest and Timber Inquiry: Final Report*, Australian Government Publishing Service, Canberra.

Ruben, E.S. (1983) 'International Pollution Control Costs of Coal-Fired Power Plants', *Environmental Science and Technology*, 17, 8:366A-377A.

Sinden, J.A. and A. Worrell (1979) *Unpriced Values: Decisions Without Market Prices*, Wiley, New York.

Starkie, D.N.M. and D.M. Johnson (1975) *The Economic Value of Peace and Quiet*, D.C. Heath, Lexington Mass.

Wilks, L. (1990) *A Survey of the Contingent Valuation Method*, Research Paper No 2. Resource Assessment Commission, Canberra.

Yang, E.J., R.C. Dower and M. Menefee (1984) *The Use of Economic Analysis in Valuing Natural Resource Damages*, NOAA, US Department of Commerce, Washington D.C.

Chapter 4

ECONOMIC IMPACT ANALYSIS

Secondary economic and financial impacts of development proposals can occur in a regional or national context. They may have major economic effects on particular regions or groups within society. The economic significance of such impacts should be interpreted with great care.

Some effects may need to be incorporated in the benefit-cost analysis of the proposed development. For example, the alleviation of chronic regional unemployment may generate economic benefits that could be included in the category of "other benefits" in the general formula for net benefits.

Assessing impacts could also be significant from an equity viewpoint. The question of who gains and who loses from development programmes and projects is, in practice, usually an important policy consideration. Assessing the distribution of effects is known as incidence analysis. Secondary economic impacts may be one element of such an analysis. Incidence effects include fiscal impacts.

Use of Input-Output Analysis

Conceptual Basis

Secondary impacts of development are usually assessed through the use of input-output models. Conceptually, input-output analysis is rather simple (Dorfman et al. 1958; Leontief 1986). The economic system is divided into separate producing sectors. Each sector sells its output to other sectors and to final buyers such as consumers, governments and export buyers. Such demands

D. James, The Application of Economic Techniques in Environmental Impact Assessment, 97–109.

are "final" in the sense that goods and services supplied to these categories pass out of the production system. At the same time, each sector buys goods and services from other producing sectors, as well as labour, land and capital inputs, to carry out production.

In order to meet final demands for goods and services, sectors have to produce enough to supply inputs to other producers as well as satisfying the final demands themselves. Secondary economic impacts are associated with this direct and indirect production response. Consumption patterns in the region/nation may also enter the assessments.

An example of economic impact analysis based on input-output models is given in Chapter 10 which describes the regional effects of a new aluminium smelter in the Hunter Valley, New South Wales, Australia. The impacts of development are assessed in terms of income and employment. Two different phases of impact are distinguished: the construction phase; and the ongoing normal operating phase.

Transactions Table

The usual starting point for the construction of an input-output model is a data set for a given base year, showing the financial transactions that occurred among the various producing sectors and final buyers within a region. These transactions are usually measured in monetary units.

A simple hypothetical transactions table is shown in Table 4.1 for a regional economy. There are only three sectors: agriculture, services and manufacturing. Financial flows are measured in dollars. (In practice, flows are usually measured in \$'000 or \$ million.)

Each sector sells to the other sectors and to final markets (households, government and exports). Inputs to production are purchased from other sectors. In addition, each sector buys primary inputs in the form of household labour, managerial skills and capital services, and imported goods and services. Note that, for each sector, the row and column totals are equal.

Direct Requirements Coefficients

To convert the transactions table into an input-output model, it is assumed that all inputs are used in fixed proportions. The per-dollar input requirements for each sector are determined by "direct requirements" coefficients, which are derived by dividing each entry in the transactions table by the corresponding column total. The table of direct requirements coefficients for the model presented here is shown in Table 4.2. It covers primary inputs and employment as well as inputs from other sectors.

Table 4.1.
Transactions Table for Three-Sector Input-Output Model

	Agr.	Serv.	Man.	H'Hold Exp.	Govt. Exp.	Exports	Total Supply
Agriculture		140	160	900	500	700	2,400
Services	720		320	360			1,400
Manufacturing	480	140		180			800
Household Income	768	840	192				1,800
Other Income	192	280	48				520
Imports	240		80				320
Total Outlays	2,400	1,400	800	1,440	500	700	7,240
Employment	180	140	30				350

Table 4.2.
Direct Requirements Coefficients for Three-Sector .
Model Derived from Table 4.1

	Agr.	Serv.	Man.
Agriculture		0.10	0.20
Services	0.30		0.40
Manufacturing	0.20	0.10	
Household Income	0.32	0.60	0.24
Other Income	0.08	0.20	0.06
Imports	0.10		0.10
Total Outlays	1.00	1.00	1.00
Employment	0.075	0.100	0.0375

Open Model

An "open" input-output model is used to estimate, for each sector, the level of output required directly and indirectly to satisfy a given set of final demands.

This involves solving a set of simultaneous equations, as set out below. It is assumed that, in general, the regional economy has *n* sectors.

$$x_1 = a_{11}x_1 + a_{12}x_2 + \dots + a_{1n}x_{1n} + f_1$$
$$x_2 = a_{21}x_1 + a_{22}x_2 + \dots + a_{2n}x_2 + f_2$$
$$.$$
$$.$$
$$x_n = a_{n1}x_1 + a_{n2}x_2 + \dots + a_{nn}x_{nn} + f_n$$

where $x_1, x_2 \dots, x_n$ represent sectoral outputs, a_{ij} are the direct requirements coefficients indicating input of the *i*th sector's output required to produce one unit of the *j*th sector's output, and $f_1, f_2, \dots, f_n$ are the final demands to be satisfied. This version of an input-output model is known as an "open system".

A set of coefficients known as "Leontief inverse" or "total requirements" coefficients can be derived that will indicate the direct and indirect effects of final demands on output levels (James and Throsby 1973). The total requirements coefficients for the worked example here are shown in Table 4.3.

Table 4.3.
Leontief Inverse Coefficients for Three-Sector Model
Derived from Table 4.2

	Agr.	Serv.	Man.
Agriculture	1.0959	0.1370	0.2740
Services	0.4338	1.0959	0.5251
Manufacturing	0.2626	0.1370	1.1073
Output Multiplier	1.7923	1.3699	1.9064

Denoting the Leontief inverse coefficients by a set of c_{ij}, the general set of equations can be written as follows:

$$x_1 = c_{11}f_1 + c_{12}f_2 + \dots + c_{1n}f_n$$
$$x_2 = c_{21}f_1 + c_{22}f_2 + \dots + c_{2n}f_n$$
$$.$$
$$.$$
$$x_n = c_{n1}f_1 + c_{n2}f_2 + \dots + c_{nn}f_n$$

Given a set of final demands, it is possible to solve these simultaneous equations to determine the equilibrium set of output levels.

Closed Model

A more complicated version of the input-output model, known as a "closed model", can be developed by classifying households as an additional sector (Robinson 1972). A set of household consumption expenditure coefficients is added as a new column to the table of direct requirements coefficients and household income coefficients are added to the rows. For the example given here, the closed model has four rows and columns. The direct requirements coefficients are shown in Table 4.4.

Table 4.4.
Direct Requirements Coefficients for Augmented Model
Derived from Table 4.1

	Agr.	Serv.	Man.	H'Holds
Agriculture		0.10	0.20	0.50
Services	0.30		0.40	0.20
Manufacturing	0.20	0.10		0.10
Households	0.32	0.60	0.24	

Final demands still determine the level of outputs (and total household income) for each sector. Household expenditure, however, is excluded from the set of final demands; instead it is determined within the model. The "closed" input-output model predicts the "induced effects" of consumption expenditure within the region on output levels, as well as the direct and indirect production effects. The set of coefficients for the closed model, corresponding to those of the open model, is shown in Table 4.5.

Table 4.5.
Leontief Inverse Coefficients for Augmented Model
Derived from Table 4.4

	Agr.	Serv.	Man.	H'Holds
Agriculture	2.0000	1.1220	1.1707	1.3415
Services	1.1667	1.8943	1.2520	1.0874
Manufacturing	0.6667	0.5772	1.5081	0.5996
Households	1.5000	1.6341	1.4878	2.2256

Economic Impact Multipliers

Concept of Multipliers

To predict secondary economic impacts of new development within a region, economic impact multipliers are used (Jensen et al. 1979). These can take three forms: output multipliers; Type I multipliers; and Type II multipliers. They indicate the region-wide effects of changes in final demand on output, income and employment.

Suppose the proposed development is a project falling within the same industry or sectoral category as an existing industry in the region. If the project has similar input requirements and sells to the same regional markets as the existing industry, the multipliers associated with its establishment will be the same as those for the industry or sector to which the project belongs. Impacts of general sectoral expansion programmes can likewise be assessed with multipliers yielded by the input-output model.

However, if the proposed development involves projects or sectors that are new to the region or rely on production technology that is different from existing technology, new rows and columns must be added to the input-output table to account for structural change in the region. A new input-output model must be built.

Secondary impacts may vary over time. For example, introduction of a major project may generate jobs for construction workers and supporting services during the capital installation phase, but have an entirely different set of effects once the project becomes operational. Separate input-output studies for the different phases are required to determine the temporal distribution of impacts. At the regional development level, the technique may be usefully applied to manage more effectively the "boomtown" effect.

Output Multipliers

Total output multipliers for the three-sector model are shown in Table 4.3. The column sum of the Leontief coefficients for the *j*th sector measures the increase in output occurring in all sectors as a result of a one-unit increase in final demand for the *j*th sector. An extra dollar's worth of final sales by the agricultural sector will lead to an increase in output of $1.7923 for the whole region. For services and manufacturing, the corresponding total output multipliers are 1.3699 and 1.9064 respectively.

Income and Employment Multipliers

The regional effects on income and employment of change in final demands associated with new development proposals can be assessed by means of Type I and Type II multipliers. Type I multipliers are derived from the open input-output model; they identify changes in regional income and employment generated directly and indirectly through production linkages in the region. Type II multipliers are derived from the closed model; they include consumption-induced changes in income and employment as well as direct and indirect production effects. Such multipliers are shown in Table 4.6.

Table 4.6.
Type I and II Multipliers for Income and Employment
in Three-Sector Model

	Agriculture	Services	Manufacturing
HOUSEHOLD INCOME			
Direct coefficient	0.3200	0.6000	0.2400
Direct and indirect	0.6740	0.7342	0.6685
Direct, indirect and induced	1.5000	1.6342	1.4878
Type I multiplier	2.1063	1.2237	2.7854
Type II multiplier	4.6875	2.7237	6.1992
OTHER INCOME			
Direct coefficient	0.0800	0.2000	0.0600
Direct and indirect	0.1902	0.2384	0.1934
Direct, indirect and induced	0.4333	0.5033	0.4346
Type I multiplier	2.3775	1.1920	3.2233
Type II multiplier	5.4164	2.5165	7.2433
EMPLOYMENT			
Direct coefficient	0.0750	0.1000	0.0375
Direct and indirect	0.1354	0.1250	0.1146
Direct, indirect and induced	0.2917	0.2952	0.2696
Type I multiplier	1.8053	1.2500	3.0560
Type II multiplier	3.8893	2.9520	7.1893

The Type I employment multiplier for each sector is the ratio of the direct plus indirect employment coefficient to the direct employment coefficient. Each Type I employment multiplier indicates the additional employment that will be generated directly and indirectly throughout the region for each new job created within a particular sector by an increase in final demand for that sector's output. For example, the Type I employment multiplier for manufacturing is 3.056. This means that an increase in final demand creating a new job in manufacturing will create an additional 2.056 jobs in other sectors, bringing the regional employment effect to 3.056.

Type II multipliers are derived in a similar way, using the closed input-output model. Each Type II multiplier measures the total amount of employment generated directly, indirectly and induced, for each new job created by an increase in non-household final demand within each sector.

Type I and Type II multipliers for household income and other income can be obtained, using similar calculations. These multipliers measure the total amount of extra income that is generated for each $1 change in income resulting from a change in final demand for a particular sector.

The complete set of Type I and Type II income and employment multipliers for the hypothetical model developed here is presented in Table 4.6.

Interpretation of Secondary Economic Impacts

While information yielded by an input-output study on the incidence of economic impacts of development may be taken into consideration in decisions on development proposals, quite a different interpretation of results must be made from an economic efficiency perspective. From this perspective, results of input-output calculations conducted in EIAs are often poorly interpreted. For example, it is commonly assumed that all jobs and income predicted to flow from a new development project represents an economic benefit to the community. It may also be assumed that if a project did not go ahead, or if an existing project were to be closed down, the jobs and income calculated by the input-output model would represent an economic loss. There is a basic misconception in this assessment.

The key to correct interpretation, at least in terms of social gains and losses, must be found in the "with" and "without" situations. In a benefit-cost analysis, jobs in themselves have no meaning. It is only the effect on the net income of the community that matters. This depends on the shadow price of labour. The direct employment of labour in a development project usually represents a cost to the community, not a benefit, because it involves the use of a scarce resource with alternative uses. In a full employment situation, labour will merely be transferred to the project from other production activities, with no net effect on the community's real income.

As regards labour indirectly employed, the input-output calculations show only an association of jobs and income with the project. What must be predicted is what would happen in a region if the project did not go ahead. Alternative employment and income-earning opportunities are affected by a wide range of factors such as the general state of the economy, alternative development proposals and the economic and geographic mobility of the work force. If labour simply moves from one set of activities to another, the net effect on jobs and employment may be zero.

This argument extends to labour employed by the government sector to support development projects. Such secondary impacts are often claimed as a benefit in an EIA. In fact, they are more likely to represent a cost to the community or taxpayer.

In situations where a pool of unemployed labour exists, the scarcity price of labour or the shadow wage rate is much less than the market wage rate. From a production viewpoint, the marginal opportunity cost of surplus labour is zero. However, to induce labour to enter the work force, a price must be paid that is high enough to compensate for the loss of leisure time. This price is the shadow wage rate that should be applied in a benefit-cost analysis.

Applications to Cumulative Environmental Impacts

Secondary impacts of projects on other industries in a regional economy can result in significant indirect sources of pollution and other adverse environmental effects. Many "clean" industries can have potentially disastrous environmental impacts because of the feedback effects on environmentally damaging supplier industries. These indirect or region-wide effects can be predicted through the construction and use of "cumulative" pollution coefficients. Space does not permit an exposition of the relevant modelling techniques here; they are described fully by James et al. (1978) and Hufschmidt et al. (1983) among others. The application of cumulative pollution coefficients in regional development planning and environmental impact assessment is demonstrated in Chapter 10.

Practical Aspects of Input-Output Modelling

Input-output models are a useful means of assessing the secondary economic impacts of development proposals, as well as indirect regional environmental impacts. The models do, however, have certain limitations. Many economists strongly criticise the assumption of fixed coefficients which implies simple linear relationships among all variables. In reality, economic functions are frequently nonlinear. Furthermore, relationships change over time. The coefficients in an input-output model are valid only for the base year. For later years, new coefficients might have to be used. The models are also based on an assumption of non-substitution of inputs in production. Over time, and particularly as a result of changes in technology, costs and prices, substitutions do tend to occur.

Another problem with input-output models is the difficulty of obtaining base-year data. All producers and consumers in a region must be surveyed. This is usually a time and labour consuming task. There are various short-cut methods of compiling input-output tables, but they often have theoretical and practical limitations (Richardson 1972; Jensen et al. 1979).

Although the basic concept of input-output models is simple, a fair degree of mathematical ability is required to construct and apply the models. It is easy for an unwary analyst to perform the necessary calculations incorrectly or interpret the results poorly. Computing skills are essential. Most calculations can be performed on a personal computer. An operational regional model would typically have twenty to thirty sectors. Larger models of the macro-economy might feature more than a hundred sectors.

Fiscal Impact Assessment

Types of Fiscal Impact

The purpose of fiscal impact assessment, or cost revenue analysis as it is sometimes known, is to predict the changes in local or state government expenditures and revenues that may occur as a consequence of a proposed development. An example of such analysis is given in Chapter 10. Another interesting case study has been documented by Krutilla et al. (1978) on the fiscal impacts of coal development in Montana.

Fiscal impacts may be divided into direct and secondary effects. Direct revenue effects refer to the additional taxes and government charges paid by the proposed development. Direct costs are mainly those of government support services. Secondary effects include government revenues from new residents, industries and commercial services attracted to the area by the development and the costs to government of supporting the additional activity.

Revenue Estimation Methods

Fiscal revenues consist of taxes (eg property tax, local income tax or sales tax), charges and inter-governmental transfers. Leistritz and Murdock (1981) describe various components of government revenue and relevant estimation methods:

Property Tax Estimation. This is based on changes in commercial and residential property and the tax base that is applicable.

Income and Sales Tax Estimation. This focuses on projected income and expenditure in the region and the taxes that would be paid. Input-output models may be applied to this task.

Production and Severance Taxes. Taxes on production may be an important source of revenue, for example charges on minerals output. Estimates can be obtained from production estimates and taxation or royalty rates.

Population-Related Taxes and Charges. These charges and taxes include fees, permits and user charges for infrastructure services such as water, sewerage and sanitation services.

Inter-Governmental Transfers. Allocation formulae for inter-governmental transfers are often complicated, so this revenue source may be difficult to project. Relevant examples include funds for education paid by a federal government to state or local authorities.

Expenditure Estimation Methods

Expenditure estimation methods identifed by Leistritz and Murdock (1981) include the following:

Per Capita Expenditures Method. This method assumes that government expenditure on each resident can be estimated as an average cost. Projection of the population thus yields an estimate of total outlays. An alternative approach is to derive a per capita multiplier through regression analysis of government expenditures, using data from a number of different regions.

Service Standard Method. This method uses estimates of the necessary service categories, equipment and capital facility requirements for different levels of the population. Cost estimates can be derived from these data.

Cross-Sectional Regression Methods. Under this approach, government services required to support various populations in different communities are estimated by regression equations with explanatory variables such as income and levels of education. Projection of the explanatory variables yields estimates of required government expenditure.

The Case Study Method. This method relies on interviews with government officials to obtain estimates of projected expenditure, based on past experience. A weakness of the case study approach is difficulty in replication.

The Comparable City Method. Data are obtained on cities of comparable size and attributes. The approach is useful in associating government expenditures with different economies of scale in city developments.

The Economic-Engineering Method. This method involves estimation of the inputs required to support infrastructure development and the provision of services, based on an engineering production function. Inputs can be costed to derive estimates of government expenditure.

Inter-Temporal Distribution of Revenue and Expenditure

The inter-temporal distribution of government revenues and expenditures depends on the timing of the various phases of the project cycle. The capital construction phase usually involves a different set of inputs and secondary economic impacts to the operational phase. Fiscal impacts need to be estimated according to the direct and secondary patterns of economic activity associated with each phase.

Local and state governments may need to cope with the problems of financing "up-front" infrastructure capital. For some developments, local governments may be able to appropriate a front-end payment to cover the cost of local infrastructure.

Allowance may also have to be made by governments for adjustment assistance packages and unemployment support if labour is displaced in moving from one phase of development to the next. Building and construction workers in particular are prone to displacement once the construction phase has been completed.

References

Dorfman, R., P.A. Samuelson and R.M. Solow (1958) *Linear Programming and Economic Analysis*, McGraw Hill, New York.

Hufschmidt, M.M., D.E. James, A.D. Meister, B.T. Bower and J.A. Dixon (1983) *Environment, Natural Systems and Development: An Economic Valuation Guide*, Johns Hopkins University Press, Baltimore.

James, D.E., H.M.A. Jansen and J.B. Opschoor (1978) *Economic Approaches to Environmental Problems*, Elsevier Scientific, Amsterdam.

James, D.E. and C.D. Throsby (1973) Introduction to Quantitative Methods in Economics, Wiley, Sydney.

Jensen, R.C., T.D. Mandeville and N.D. Karunaratne (1979) *Regional Economic Planning: Generation of Regional Input-Output Analysis*, Croom Helm, London.

Krutilla, J.V., A.C. Fisher and R.E. Price (1978) *Economic and Fiscal Impacts of Coal Development*, Johns Hopkins University Press, Baltimore.

Leistritz, F. and S. Murdock (1981) *The Socioeconomic Impact of Resource Development: Methods for Assessment*, Westview, Boulder.

Leontief, W. (1986) *Input-Output Economics*, Second Edition, Oxford University Press, New York.

Richardson, H.W. (1972) *Input-Output and Regional Economics*, Weidenfeld and Nicolson, London.

Robinson, J.N. (1972) *Planning and Forecasting Techniques: An Introduction to Macroeconomic Applications*, Weidenfeld and Nicholson, London.

Chapter 5

WATER POLLUTION IN JAKARTA BAY

ALISON GILBERT AND DAVID JAMES

This case study demonstrates the application of economic analysis to the problem of water pollution management in Jakarta Bay, Indonesia. In an existing situation where pollution damage is widespread, as has occurred in Jakarta Bay, introduction of a pollution control programme can bring significant economic benefits to the community. Such benefits consist largely of damage costs avoided. They include benefits resulting from the restoration of sustainable yields from natural ecosystems, such as aquaculture and fisheries, as well as improvements in environmental amenity values of tourism and recreation activities.

The case study does not attempt to provide actual estimates of benefits and costs, because of limitations in information, time and research resources.

Instead, it takes the form of a scoping study. It provides a good example of the kinds of information gathering and conceptualisation that should be undertaken before embarking on a detailed analysis of management options for pollution control and empirical measurement of benefits and costs.

The first part of the study comprises a preliminary scientific assessment of water pollution problems in the Bay, covering the major types of pollutants, sources, dispersion, assimilation, accumulation and environmental impacts. The study reveals large gaps in scientific information. Sufficient data are available, however, to provide general guidelines for a pollution management programme.

D. James, The Application of Economic Techniques in Environmental Impact Assessment, 111–142.

An economic framework for the assessment of options is advocated. Finally, the case study describes the legal and institutional setting within which a pollution management programme must be established and administered.

The case study draws on papers presented at the Workshop on Coastal Management and the Bay of Jakarta held in Jakarta in May 1987, organised by the Centre for Research of Human Resources and the Environment from the University of Indonesia, and co-sponsored by the Indonesian Ministry of Population and Environment (KLH) and the International Federation of Institutes for Advanced Study (IFIAS). Material is also included from a report prepared by Ecoservices Pty Ltd for the Environmental Sector Review of Indonesia conducted by the United Nations Development Programme in collaboration with KLH in 1988.

Problem Definition

The Jakarta region is a major centre of population and industrial and commercial activity. It relies heavily on its natural resource base for water supplies, fishing, agriculture, aquaculture, tourism and recreation. Its rivers and coastal areas have become a sink for the disposal of human and industrial wastes. The assimilative capacity of the environment has been exceeded, resulting in extensive degradation with accompanying economic losses and adverse community impacts. Severe problems exist with respect to heavy metal pollution, sedimentation, and pollution from organic wastes and pathogenic organisms. Substantial economic benefits would accrue from a water pollution control programme. Practical methods of measuring such gains are described in this case study.

Political and Institutional Setting

(Ben Boer and Angus Martyn)

The Republic of Indonesia consists of some 13,000 islands and islets of which less than half are inhabited. The population of Indonesia is 185 million people, with an average population density of a little under 100 persons per sq km. The inner islands of Java, Madura, and Bali which represent some 7% of the total land area, are occupied by more than three-fifths of the total population. The serious environmental problems created by the concentration of the population on these islands have been well documented for over two decades (see Danusaputro 1978).

Indonesia became an independent republic in 1945 with a unitary system of government. Under the Constitution, the sovereignty of the state is vested in the people and fully exercised by the Peoples Consultative Assembly or Majelis. The Majelis, which meets every five years, functions to sanction the constitution and its amendments, to promulgate the broad outlines of the state policy for every five years, to designate the President and Chief Executive of government, and to control the execution and implementation of the five-year state policy. An elected House of Representatives is responsible for the enactment of legislation and other activities of a more day to day nature. The country is divided into 27 provinces with a governor and regional house of representatives, each province being divided into districts, of which there are some 290.

Indonesia has a vast natural resource base including oil, natural gas, forests and fisheries and, generally speaking, plentiful water resources. Soil fertility is extremely high in Java but this is not the case in most areas of the large outer islands. Indonesia has become a major oil and natural gas exporter. The agricultural sector employs almost half of the workforce, but was replaced in the 1990s by the manufacturing and industrial sector as the largest contributor to national income. Since 1985 a number of significant initiatives have been undertaken by the government to free-up the regulatory framework for both foreign and domestic investment. This framework is nevertheless still relatively complex and further deregulation is required if Indonesia is to match the dynamic economies of neighbouring countries (Swartz 1992).

It has been realised that environmental problems are inextricably linked to population pressures, particularly in relation to waste management, water pollution and demand for both urban and agricultural land. Poverty is a major problem, especially due to deforestation by poor, marginalised farmers. Development policy, as outlined by the current (1989-94) five year development plan, is now based on a more integrated approach to ecological, economic, social and cultural issues. However, the lack of coordination mechanisms for environmental protection at the national, regional and local levels and the comparatively weak positions of environmental institutions has tended to hamper the implementation of the sustainable development imperative.

Use of the Natural Environment in the Jakarta Region

The Physical Environment

The Jakarta region is located on a coastal plain with eighteen major rivers feeding into an adjacent bay (Jakarta Bay). The total catchment area is approximately 9,000 sq km. Jakarta Bay contains fifteen coral islands and extensive coral reefs. A map of the region is shown in Figure 5.1.

Precipitation follows a monsoonal pattern. The western monsoon is accompanied by a rainy season from December to March, and the eastern monsoon by a dry season from May to October. The two monsoonal periods have different impacts on the natural environment affecting winds, coastal currents, sedimentation processes and ecological systems (Ongkosongo et al. 1987).

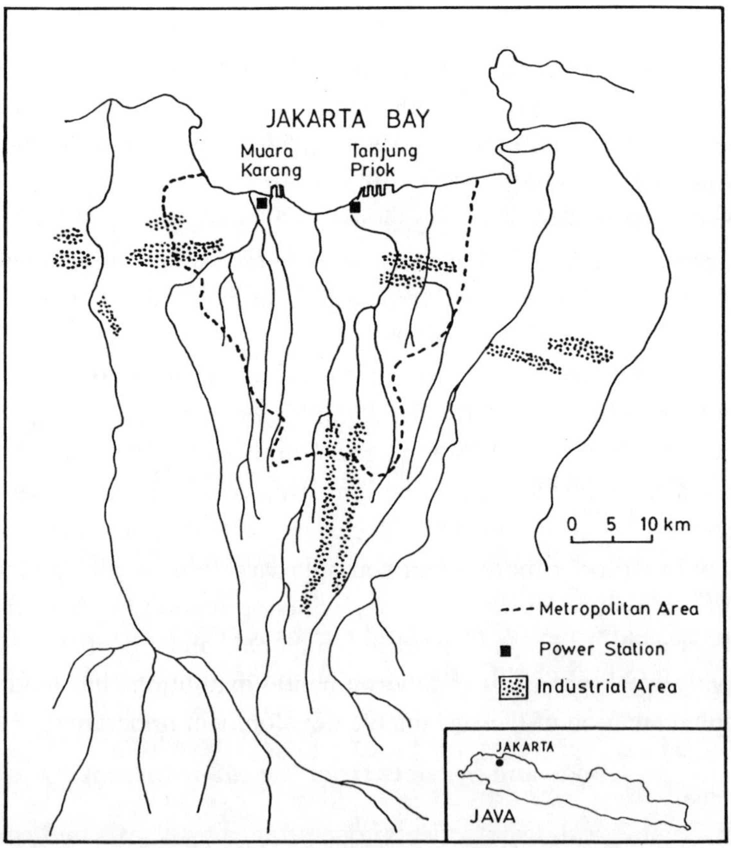

Figure 5.1. Map of Jakarta and Jakarta Bay.

Activities in the Region

The population of the Jakarta region is 11 to 12 million people. In 1948 it was only 120,000. The population of the city itself exceeds 7 million and is spread over 655 sq km in the central part of the coastal plain. Village settlements, accommodating many families dependent on artisanal fishing, occur on the coastal fringes, especially the eastern flank of the Bay.

The rapid population growth of Jakarta coincided with industrial development. Consequent demands for land and housing have stimulated land reclamation, mostly on the eastern flank. These demands often conflict with other land uses, particularly aquaculture and preservation of natural shore environments such as mangrove communities.

There are more than 2,000 industrial plants in the Jakarta region, mostly concentrated in the JABOTABEK area in the lower part of the Bay's catchment. Metal fabrication and electroplating plants are scattered throughout the region. Other activities include automotive stamping and assembly plants, shipbuilding and repair, battery making, leather production, textiles, pulp and paper, plastics, chemicals, and food processing (Suprianto et al. 1987a).

Jakarta is a major port for domestic and international shipping. There are two main harbours with general berthing and loading facilities, as well as harbours specifically designed for the handling of fish and forestry products. No facilities exist for the disposal of bilge water or other wastes from ships.

The bay is a popular area for tourism and recreation. Activities include boating, fishing, and picnicking, although fishing is becoming less popular due to fears of contamination. Swimming is largely precluded because of water pollution. On the coral islands, swimming, skindiving, fishing and boating are possible. Prospects for an expansion of tourism and recreation will be greatly enchanced with declaration of Kepulauan Seribu Marine National Park (Suprianto et al. 1987b).

Many activities in the Jakarta region are dependent on river water. Water uses include irrigation, salinity control in aquaculture, process and cooling water in industry, and water for household purposes. The rivers also serve as a sink for the disposal of household, urban, agricultural and industrial waste.

Goods and Services from the Environment

Fisheries. Commercial fishing fleets, operating mostly offshore, market their catches in Jakarta. Local fishing is conducted by artisanal fishermen using traditional methods. Artisanal fishermen in the Jakarta region once fished primarily in the Bay. They have been now forced by increased demand, and environmental degradation, particularly of the coral communities, to travel greater distances and cover larger areas. These fishermen have invested heavily in motorised boats. Fish are still caught from illegal fishing platforms in the Bay.

The catch marketed in Jakarta in 1985 was 20,000 tons of marine fish, valued at Rp 12.5 million. This figure does not distinguish between fish caught locally and those from outside sources, nor does it account for the informal markets that are almost certain to exist. Catch statistics show an increasing trend between 1968 and 1980, a sharp decline in 1981, and a very gradual decline since. The possible causes of this decline include reduced fish stocks, government intervention, and the rise of informal markets (Wahyuningsih 1987).

The potential for recreational fishing in Jakarta Bay has been severely diminished by poor environmental quality, reduced fish stocks, and contamination. Potential still exists around the few coral reefs that retain healthy communities.

Aquaculture. Indonesia has a long tradition in the culture of brackish water organisms in tambaks (ponds). Juveniles are captured in the wild and grown in tambaks along the coast. The ponds are filled and flushed with water from the Bay to maintain water quality, and diluted with good quality freshwater when necessary. Important species include bandeng or milkfish (*Chanos chanos*), mullet (belanak), seabass (kakap), and tilapia (mujair). Shrimps may be grown with milkfish as a secondary crop. Tiger shrimp or udang windu (*Penaeus monodon*) has become increasingly important in the last decade as a result of rapidly growing world markets. Aquaculture near Jakarta has declined by more than 50% since 1971. Two likely causes are the annexation of tambaks for urban development, and the pollution of freshwater which is necessary for salinity control (Risch 1987).

Nature Conservation. Conservation along the shores of Jakarta Bay has involved retention and restoration of mangrove communities at three sites representing a total of 1140 ha. Declaration of these reserves is formal recognition of the role of mangroves in coastal protection, pollution recycling, marine nurseries and as an energy resource. Demands for aquaculture and urban development tend to take precedence over resource conservation, resulting in continued mangrove reclamation, even in the declared reserves.

Two areas immediately offshore have been declared nature reserves: Palau Rambut (45 ha) and Palau Bokor (15 ha). Both islands support a large number of migratory and local sea birds, and are associated with coral reefs where ornamental and commercial fish species may be found.

Attempts have been made to protect certain marine species. The two protection instruments used are the listing and protection by legislation of some endangered species, and the protection of habitat. As a signatury to the Convention of International trade in Endangered Species (CITES), Indonesia is obligated to take measures to prevent smuggling of protected species.

Sand and Coral Mining. Rapid growth of the City of Jakarta began in the 1950s. The demand for construction sand was met by extraction from the shores of the Bay. This initially small industry grew into a large unregulated industry. The government has made a number of attempts at control and regulation, but illegal sand mining still continues. According to the Ministry of Forestry, 2000 tonnes of sand were being transported every day in 1984 from the Bay to Jakarta. The sand is taken primarily from beaches on the eastern shores of the Bay. These beaches are naturally accreting as a result of heavy sediment loads in the rivers. Accretion rates vary from 40-50 m/yr in the river mouth to a maximum of 9 m/yr on the beaches. On Ancol Beach and near Marunda, sand mining is resulting in net erosion of 5m/yr and 14 m/yr, respectively.

High sedimentation rates have led to the need for dredging, particularly to keep channels open for traffic entering the harbours of Tanjung Priok and Sunda Kelapa. Between 150,000 and 200,000 cu m of sediment are dredged annually from the western navigation channel of Tanjung Priok alone. Some dredge spoil has been used for reclamation purposes and flood mitigation projects.

Mining of coral and white sand occurs, with government approval, on a number of islands in Jakarta Bay and in the Kepulauan Seribu. White sand mining has an estimated capacity of 10,000 cu m/day.

Oil Exploration. Oil exploration and production occur in the Java Sea. Two companies, IIAPCO Inc. and ARII Inc., are authorised to conduct explorations near the Kepulauan Seribu, in a joint venture with Pertamina, the Indonesian Government's gas and oil company. Oil tankers in the shipping lanes offshore from the Kepulauan Seribu have been reported to flush their holds and to discharge oily water into the sea. Tar balls and oil residues have been found on the beaches. This is of some concern, given the proximity of IIAPCO to the Kepulauan Seribu Marine National Park.

Electricity Generation. Two power stations are located on Jakarta Bay: Muara Karang PLTU and PLTU Tanjung Priok. Both are oil-fired and discharge cooling water into the Bay. Muara Karang PLTU has a capacity of 700 MW. PLTU Tanjung Priok has a total capacity of about 150 MW, although only one of the four units (25 MW) is actually in operation. For both plants, the temperature of intake water is about 32°C, and outlet water about 40°C, a difference of 8°C.

Waste Disposal. The canals and rivers flowing into Jakarta Bay, as well as the Bay itself, play a major role in waste disposal in Jakarta. The nature and extent of the overloading of environmental assimilative capacities are discussed in detail below.

Conflicts in Resource Use

A number of conflicts exist among the various activities in the Jakarta region. This case study focuses on conflicts between the use of Jakarta Bay for waste disposal and other activities dependent on the natural environment.

Fisheries. Some commercial fish species spawn in open water but spend much of their early life in estuaries and other inshore areas. Juvenile mortality in such species probably increases as a direct result of pollution in these nursery areas. This, in turn, adversely affects adult populations. Heavy metals and organic micro-pollutants are especially damaging. Juveniles are more exposed to these substances and are more sensitive to them. Sub-lethal doses may result in bioaccumulation.

Oil pollution has the potential to affect fish at all stages of life, since spills occur in harbours, inshore areas and at sea. The toxic effect of the volatile components is probably greatest on the very young, planktonic stages, as older individuals are sufficiently mobile to avoid most spills.

Shoreline development causes habitat modification, including the nursery areas of mangrove and seagrass communities. Habitat modification is also an indirect result of pollution. Anoxia, resulting from high nutrient loads has caused widespread loss of estuarine habitat. Oil pollution, notably the settling of oil residues, has been responsible for changes in habitat, species composition and disruption of marine food chains.

Concentrations of heavy metals in fish taken from Jakarta Bay and nearby areas commonly exceed current standards, especially for cadmium, lead and mercury. The high levels of these contaminants are a matter of serious concern. It would appear that contamination is not confined to the immediate vicinity of the Bay.

Contamination of commercial species by toxic substances is a major threat to the whole community. Fishing communities are at greatest risk, as controls on their consumption of contaminated products are almost impossible to implement.

Sand and Coral Mining. Mining activities contribute to sediment loads, adversely affecting coral communities, mangrove stands and seagrass beds. Corals can be choked by settled materials and their metabolism affected by reduced water clarity. Sedimentation can damage mangroves by smothering pneumatophores. Dredging activities resuspend particles containing heavy metals or organic micro-pollutants, increasing their bioavailability. Contaminated dredge spoil used for land reclamation may also have adverse environmental impacts.

Reclamation near Pantai Mutiara, adjacent to the power plant at Muara Karang, is reducing the volume of water into which the plant discharges its cooling water. This reduces heat dispersion, adding to thermal stresses on the environment, as well as increasing the temperature of intake water. The latter can be expected to reduce the life of the plant.

Oil Drilling and Exploration. The main risk from oil activities is the discharge of oil from tankers, both at Pabelokan island, just outside the Marine National Park, and in the Bay itself. The visual qualities of parts of the Marine National Park have already been adversely affected with reports of oil residues occurring on beaches.

Analysis of Pollution Problems

Major Pollutants and Their Sources

Wastes enter Jakarta Bay via a number of routes, direct and indirect. Direct disposal involves thermal wastes from power stations, solid wastes, and discharges from sea-based activities such as shipping, pipelines and oil platforms. Most wastes enter the Bay indirectly through land drainage, urban stormwater runoff, or atmospheric deposition.

Land drainage carries pollutants from point and diffuse sources. Point sources include direct discharges by industry and residences into rivers and canals. Diffuse sources are typically associated with agricultural activities and result in discharges of pesticides, nitrates, phosphates, solid wastes and organic matter being transported into the river system by surface and sub-surface runoff. Sediment loads have been increased as a consequence of agricultural activities, especially in the upper catchment.

Aquaculture is contributing to organic and chemical residue levels, following the move towards more intensive aquaculture and the development of tiger shrimp cultivation.

Urban storm water runoff carries a variety of dissolved, suspended and solid wastes, including lead from automobile exhausts, oils, nutrients, pesticides and domestic rubbish. Atmospheric deposition of particulates derived from combustion processes may be a source of heavy metals.

Solid waste production is estimated to have grown from 1200 cu m/day in 1976-77 to 17,940 cu m/day in 1984. Organic waste comprises 77% of the total, with paper, plastic, wood, metal and textiles ranging from 2.5% to 6.5%. The sources of these wastes are: domestic wastes 65%; market wastes 15%; industrial wastes 10%; construction, recreation and office wastes 10%. About 79% of this waste is managed by government or non-government organisations. The rest is discharged directly into the rivers and Bay (Suprianto et al. 1987a).

Approximately 1,500 cu m of sewage are generated each day. There is no sewerage reticulation system, no widespread use of septic tanks, and no sewage treatment. Large quantities of untreated sewage are discharged directly into surface water systems through open channels and drains. These sewage discharges have led to deoxygenation of streams, algal blooms, pathogenic contamination and malodour.

The main sources of industrial wastes are industries located in the JABOTABEK area, in particular the metal, battery, leather and textile industries. Of particular concern are heavy metals, particularly lead and mercury. Large quantities of organic wastes are also discharged, especially from pulp and paper plants, and food and beverage processing facilities. BOD levels up to 100 times those of domestic sewage have been recorded. Other industrial wastes include oils, phenols, PCBs and other organic micro-pollutants, cyanides and suspended solids.

Sand mining, coral mining and dredging contribute to sediment loads in Jakarta Bay. Oil and bilge water are discharged by ships. There is also a risk of oil slicks from oil and gas exploration carried out in the Bay.

Behaviour of Pollutants in the Environment

In general, the main impacts of toxic waste disposal on marine ecosystems are:

- direct toxic effects on organisms (e.g. effects of the volatile components of oil on corals);
- sublethal effects which increase stress (eg thermal discharges and heavy metal pollution);
- accumulation in sediments (e.g. oil residues, heavy metals and organic micro-pollutants);

- alteration of the chemical environment and overloading of buffering systems as occurs with acids, alkalis, magnesium and calcium;
- habitat modification and loss;
- overloading of detritus foodchains and subsequent reductions in oxygen levels;
- stimulation of plant growth which, if extreme, causes nocturnal deoxygenation;
- generation of unpleasant sights and smells.

The highest concentration of nutrients is found near river mouths, although this depends on seasonal factors. Greater runoff and discharge in the wet season results in higher nutrient loads and a more even distribution throughout the Bay. Phytoplankton and zooplankton tend to follow these distribution patterns.

Many of the canals that drain Jakarta have high loads of organic material and greatly reduced dissolved oxygen concentrations. Dissolved oxygen concentrations in the Bay itself are close to saturation levels, indicating the importance of tidal exchange and mixing by wave and current action. Measurements in the vicinity of river and canal mouths show consistently low oxygen levels.

The high loads of nutrients entering the Bay have caused enrichment and eutrophication. Eutrophication causes changes in algal species composition, adversely affecting other organisms including fish, shellfish and reef communities. Red tides involving Trichodesmiun and Noctiluca are known to have occurred (Ongkosongo et al. 1987).

Analyses of Bay water have shown excessively high levels of the bacteria Pseudomonas, Vibrios, Achromobacteria, Flavobacterium, Micrococcus and coliforms. Pathogenic bacteria have been found in samples of various organisms in the Bay. Bacterial contamination of shellfish and its health impact is a major concern.

Analyses of hydrocarbons in Bay sediments and waters indicate oil pollution in nearshore areas. Highest measurements occur in the vicinity of river mouths, harbours and immediately offshore from the City of Jakarta.

Infrared aerial photographs show the heated discharges of the power plants flowing as far as 5 km to the west and 0.54 km offshore.

Assessment of Pollution Damage

Jakarta Bay is greatly influenced by sedimentation processes. Rivers are the primary source of sedimentation, and the greatest rates of sedimentation usually occur at river mouths. Activities in the catchment, in particular the clearing of forested land for agriculture, increase erosion and the quantity of sediment carried by draining rivers.

The increase in sedimentation in the bay can be gauged by comparing the annual average sedimentation rate over 6600 years with the annual average progradation rate between 1971 and 1982. Whereas the natural progradation rate over 6600 years has been estimated at between 1.0 and 4.2 m/yr for rivers in the region, between 1971 and 1982 the rates varied from 2.2 to 72.7 m/yr depending on location.

Pollution in its various forms has had serious impacts on fisheries in the region. Observed impacts fall into five main categories:

1) Changes in food availability, mainly as a result of nutrient enrichment and eutrophication.
2) Deoxygenation due to increased oxygen demand for degradation of organic matter and algal respiration (eutrophication), particularly at night.
3) Reduced survival of young fish due to the presence of toxic substances in bay water, sediments, and food.
4) Bioaccumulation of toxic substances such as heavy metals and organic micro-compounds in fisheries products, and the potential effects on human health.
5) Contamination, particularly of shell fish, by pathogenic bacteria, with consequent impacts on human health.

The first three factors adversely affect stocks, and the last two influence the quality of fisheries products and their acceptability for human consumption. Both have important economic impacts.

The restricted mobility of cultured organisms in aquaculture increases their exposure to contaminated water and suspended sediment in fresh and bay water. Consequently, bioaccumulation is becoming an important issue for aquaculturists. The poor quality of fresh water has led to the use of groundwater reserves. There are fears that subsequent recharge with Bay water will result in the contamination of ground water supplies.

The water based recreational values of Jakarta Bay proper are limited because of pollution. Recreational and subsistence fishing, as well as the collection of shellfish for own consumption, still occur, raising fears of health problems.

Coral reefs that once existed in Jakarta Bay have died during the last 10-20 years. Smothering by sediment is probably the main cause. The number of resident and migratory birds is decreasing rapidly. Pesticide accumulation is thought to be one cause, although loss of habitat and reduced food availability are also implicated.

The accumulation of heavy metals and organic micro-compounds in the sediments of the Bay may cause future problems in the use of dredge spoil for reclamation purposes, and its disposal in general. So far, sediments are not showing exceptionally high levels of pollution.

Behaviour of Heavy Metals in Coastal Zones

The general behaviour of pollutants entering marine ecosystems is illustrated in Figure 5.2. Investigations into heavy metal pollution have been carried out by the Hydro-Chemical section of the Institute of Hydraulic Engineering. Data were collected for sources in the city and industrial regions and in the down-stream areas of the rivers. Data were also gathered on the heavy metal content and the annual discharges of inflowing rivers. The total transport of heavy

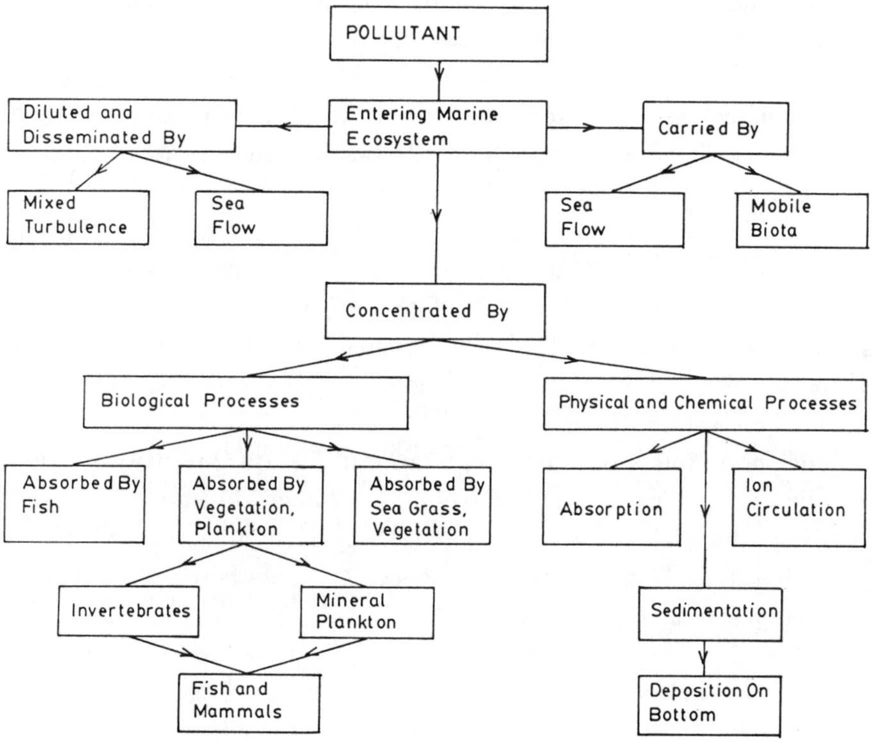

Figure 5.2. Flow chart of pollutants entering marine ecosystem. (Source: Center for Environmental Studies, Gajah Mada University (1987).

metals into Jakarta Bay has been estimated at: arsenic 45.5 kg/day; chromium 325 kg/day; copper 191 kg/day; mercury 362 kg/day; nickel 154 kg/day; lead 756 kg/day; and zinc 42,983 kg/day. Measurements of heavy metal loads are taken only five times per year, so these figures are only rough estimates.

Heavy metals are present in a number of biotic and abiotic compartments of the environment, and in various forms, each of which may have a different geochemical behaviour and bioavailability. Abiotic reserves consist of deposited sediments, interstitial (pore) water, suspended matter and surface water. Under reducing conditions, sediments release nutrients and heavy metals into interstitial water. This may then influence the concentrations in the surface water via the largely physical processes of diffusion, consolidation, erosion and bioturbation. Biologically mediated reactions also occur in sediments, such as the transformation of metal ions into methylated forms.

The contribution of abiotic reserves to the heavy metal burden of aquatic organisms depends on the metal's bioavailability and the organism's habitat. Filter feeders and planktonic organisms are exposed to both dissolved and particulate heavy metals. Deposit feeders and meiofauna are exposed to sedimentary particles and associated pore waters. All of these organisms are ingesting bacteria associated with sediments and organic material, all which may have begun the process of accumulating heavy metals.

The behaviour of heavy metals in estuarine and near shore coastal zones is complex, due to strong gradients in chemical composition of the water (especially pH, salinity, and dissolved oxygen concentrations), variable suspended matter concentrations (often a seasonal factor), and complex hydrodynamic processes associated with interactions between tides, currents, wave action and freshwater inflow.

Mercury pollution is of major importance in Jakarta Bay. Its turnover in local aquatic ecosystems is comparatively well known. In humans, alkylmercurials easily pass the blood-brain barrier and may accumulate in the brain, where they destroy selected brain cells and produce Minimata Disease. They also easily cross the placenta, causing high mercury levels in foetal red blood cells. Developing foetal nerve tissue is very sensitive to the destructive effects of alkylmercurials, and congenital neurological damage has been reported in human foetuses without alkylmercury poisoning in the mother.

Framework for Pollution Management

Economic Analysis of Options for Pollution Control

Pollution management should be part of broader regional development plans. Economic analysis of options for pollution control can be undertaken within the broader planning process. For Jakarta Bay, there are many options. Possible measures, shown in Table 5.1, can be considered individually or as combinations.

The first set of controls relates to the structure of development and land-use planning. By controlling the sectoral composition of the regional economy, discharges of pollutants can be managed. Certain catchments may be reserved for specific functions such as the supply of drinking water or the provision of industrial cooling water which may be of quite different quality. Economic activity can then be guided in the respective catchments by means of economic incentives, regulations and development controls.

To protect the environment in the Jakarta region, planning measures should be extended to the upper limits of the catchments. A possible land use pattern would be the development of forestry and non-polluting industries in the upper catchments, buffering surface flows and providing clean water for reservoirs. Freshwater fisheries and agriculture could be developed further downstream, leaving heavy industry and urban development to the middle and lower reaches of the river systems. Full advantage should be taken of the self-cleansing powers of streams, as occurs with the natural degradation of organic waste.

Other options shown in Table 5.1 include those designed to control the generation of pollutants, their composition, rates of discharge and behaviour in the environment. The exposure of receptors may also be controlled.

To assess each option for pollution management in economic terms, several tasks must be carried out (see Figure 5.3):

1) Identification and quantification of any reduction in the discharge of pollutants achieved by the option.
2) Prediction of changes in ambient concentrations of the pollutants at receptor points using environmental dispersion and reaction models. Sediment transport models are needed to predict sediment loadings and deposition. (Some options operate directly on ambient concentrations.)
3) Determination of exposure receptors (fish, shellfish, human and economic activities) to ambient concentrations. (Some options are designed to change the exposure of receptors.)

4) Estimation of the reduction in physical damage associated with the change in environmental conditions, using dose-response functions. For example, crop productivity may increase, higher fish yields may be obtained, and human health may improve.
5) Determination of economic benefits (reduction in damage costs) generated by the option, using appropriate economic valuation techniques.
6) Estimation of the economic cost of the option, using opportunity cost methods.
7) Comparison of benefits and costs and calculation of net benefits generated by the option.

Table 5.1.
Options for Water Pollution Management in Jakarta Region

General Controls
Manage structure of regional economy
Land-use planning
Relocation of dischargers
Reduction of Pollutant Generation
Change product mix
Alter input mix
Introduce new process technologies
Change crops and agricultural practices
Treatment of Effluent Discharge
Toxics/Heavy Metals
Chemical treatment (inplant)
Chemical treatment (centralised plant)
Suspended Solids
Weirs
Settling tanks
Settling ponds
Screening
Filtering
Flocculation
Coagulation

Table 5.1 Continued.

Organic Material
Biological filtration
Activated sludge plant
Oxidation pond
Anaerobic digestion
Pathogens
Chemical treatment
Dilution
Transportation and Assimilation of Waste
Aeration
Mixing
Composting
Landfill
Flushing
Dredging
Disposal by barge, road and rail
Pipeline
Ocean outfall
Reduce Exposure of Receptors
Relocation of receptors
Relocation of water intakes
Treatment of intake water
Establish buffer zones
Engineering structures - dikes, breakwaters, flow diversions

Engineers and scientists are needed to conduct the analyses of the effects of each pollution control option on rates of discharge, dispersion patterns, possible synergisms and antagonisms, and receptor responses. Economic evaluation involves translating the physical, biological and technical parameters of each option into benefits and costs.

From model simulations, a range of alternative pollution control strategies can be identified and evaluated in economic terms. The best economic option maximises the net benefits of control, such that the marginal benefits equal the

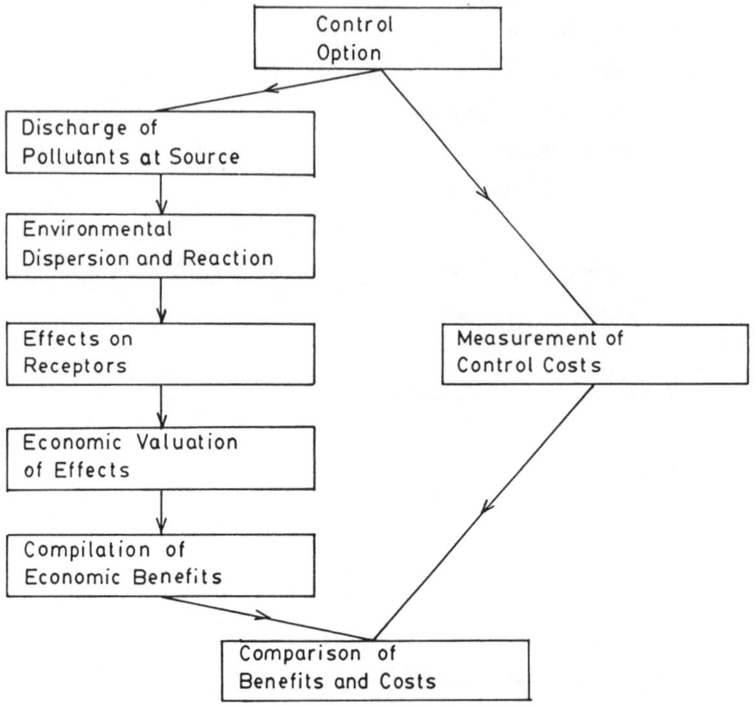

Figure 5.3. Economic framework for management of pollution.

marginal costs. In practice, the trade-off between benefits and costs of control is not continuous, but involves a number of descrete choices, each dependent on the technical, environmental, and economic parameters of each improvement option.

Difficulties in carrying out model simulations should not be underestimated. There may be a lack of scientific data or restrictions on simulation modelling capabilities. Cyclical environmental changes, such as changes in Jakarta Bay's coastal currents throughout the year, may further complicate analysis. The passage of toxic substances through the physical environment and food chains is even more complex.

Investigations of the economic implications of pollution control require analysis of the whole pollution profile as measures aimed at only single pollutants may lead to uneconomic investment in pollution control equipment and facilities, with only a neglible improvement in environmental quality.

Economic Benefits of Control

Reduced exposure of people and activities to waterborne pollutants in the Jakarta region would result in several kinds of economic benefits. A general summary of the likely benefits is given in Table 5.2. The major categories of benefits are listed, with corresponding indicators and the economic valuation techniques that may be applied. The table also indicates whether benefits are based on financial or market data, or on imputed data obtained from the use of techniques such as the travel cost method, hedonic pricing or contingent valuation.

A significant benefit in the household sector would be an improvement in health. Relevant indicators of this effect are a reduction in the number of days of illness, and a lower mortality rate. An important economic measure of such benefits is the additional income that would be earned as a result of an increase in the number of days worked. This is one application of the productivity approach as a measurement technique.

Most of this extra income would show up in national accounts as an increase in GDP, but for some types of work, which occur outside the market system, such as artisanal fishing or subsistence agriculture, the increase must be imputed.

Another benefit of improved community health is a reduction in medical expenditures. This can be measured in financial terms, or in imputed terms where medical services are provided "free". It is not possible to measure accurately in monetary terms all the benefits of an improvement in health. For example, reduced sickness in the very young and old will have no effect on the supply of labour and hence on money incomes. The reduction in suffering associated with illness and premature death also cannot be valued very easily in economic terms. Such benefits should at least be described qualitatively.

Various additional benefits would accrue to households from an improvement in the water quality on the Jakarta region. One such benefit would be a reduction in water treatment costs incurred by households for drinking, bathing, washing and cleaning. This would show up in financial terms as a reduction in outlays on fuel, soap, cleaning materials, filters and sludge disposal. Reduced maintenance, repair and replacement costs for household equipment, clothing and linen would constitute additional economic benefits. Some savings might be unpriced, such as reductions in household labour required to treat domestic water supplies. Any increase in the imputed value of homegrown garden produce or fish, as well as fish caught from rivers and Jakarta Bay directly consumed, would also comprise economic benefits.

Table 5.2.
Summary of Benefits from Improvement in Water Quality

Category	Physical Indicator	Economic Benefit	Type of Benefit	Measurement Technique
Households				
Morbidity	Increase in in working days	Increase in wages, income	Financial, imputed	Productivity approach
	Reduced medical treatment	Decrease in medical costs	Financial	Cost analysis technique
Mortality	Reduced mortality	N.A.	Qualitative	N.A.
Water Use	Reduced treatment	Decrease in treatment costs	Financial	Preventive expenditures method
	Reduced OM&R household items	Decrease in OM&R costs	Financial, imputed	Cost analysis technique
Fishing, gardening	Increased yields	Increase in value of yields	Imputed	Productivity method
Recreation	Improved amenity	Increase in consumers' surplus	Imputed	Travel cost method
Aesthetics	Improved amenity	Increase in consumers' surplus	Imputed	Contingent valuation method
Property value	Improved amenity	Increase in property value	Financial	Property value method

Table 5.2. *Continued*

Fisheries				
Finfish	Higher yields	Increase in producers' surplus	Financial	Productivity approach
Shellfish	Higher yields	Increase in producers' surplus	Financial	Productivity approach
Aqua-culture	Reduced water treatment	Decrease in treatment costs	Financial	Preventive expenditures method
Tourism				
Hotels	Increased demand for rooms	Increase in producers' surplus	Financial	Demand analysis
Transport	Increased demand for transport	Increase in producers' surplus	Financial	Demand analysis
Tourist	Increased demand for services	Increase in producers' surplus	Financial	Demand analysis
Agriculture				
Crops	Higher crop yields	Increase in producers surplus	Financial	Productivity method
Irrigation water	Reduced treatment	Decrease in treatment costs	Financial	Defensive expenditures method
Agriculture equipment	Reduced OM&R	Decrease in OM&R costs	Financial	Cost analysis technique

Table 5.2. *Continued.*

Manufacturing Industry				
Intake water	Reduced treatment	Decrease in treatment costs	Financial	Preventive expenditures method
Industrial equipment	Reduced OM&R	Decrease in OM&R costs	Financial	Cost analysis
Water Supply				
Life of reservoirs	Reduced siltation	Increase in value	Imputed	Replacement cost, value of capital
Water treatment	Reduced treatment	Decrease in treatment costs	Financial	Cost analysis technique
Flood Damage				
Reduced sediment deposition	Reduced flood risk	Decrease in risk cost	Financial, imputed	Risk cost technique
Water Transport				
Harbour, river transport	Reduced dredging operations	Decrease in dredging costs	Financial	Cost analysis technique

Another class of economic benefits accruing to households would be an improvement in recreational and environmental amenity resulting, for example, from a reduction in odour, turbidity, human wastes and bacterial pollution in water systems. These benefits would be mostly imputed. The travel cost method and contingent valuation methods could be tried, but their suitability and cost would first have to be judged.

Reduction of pollution in Jakarta Bay would generate significant economic benefits for the fishing industry. Lower exposure of marine fish populations to toxic waste, nutrients and sediments would lead to an increase in fishing yields which can be valued at market prices. Such increases in productivity would result from reduced exposure of fish populations to environmental pollution, restructuring of the food chains following a reduction in nutrients, and the creation of more favourable nursery conditions in Jakarta Bay.

Initially such benefits would consist of an increase in producers' surplus in the local fishing industry. For commercial fisheries, this would appear as increased income which can be measured fairly easily. For artisanal fisheries the imputed value of additional producers' surplus would have to be estimated.

Other economic benefits in the fisheries industry would accrue to shellfish gathering and aquaculture, associated with an increase in productivity. Again, an increase in producers' surplus would occur. It is important to recognise that productivity improvements should be estimated not just with respect to output from existing shellfish beds and aquaculture ponds, but to a projected situation with an expanded industry made possible through an improved aquatic environment. A major industry, drawing on labour in the Jakarta region, is a distinct possibility in future years.

Whether the estimated increases in producers' surplus in the fishing industry would continue to accrue to producers themselves depends to a large extent on the nature of the markets supplied by Jakarta fishermen, and the competitive structure of the local fishing industry. The increase in the supply of fish may lead to a fall in the price of fish and an increase in consumers' surplus, including a transfer of real income from producers to consumers. Total welfare, consisting of the sum of producers' and consumers' surplus, would increase. Producers supplying export markets are more likely to receive fixed prices for their products and retain the gains, unless overcrowding due to new recruitment into the industry depresses individual catches. These effects can, in principle, be studied in terms of the demand and supply curves for specific products and markets.

Improvement in water quality, especially in Jakarta Bay, would greatly enhance the prospects of tourism development. Activities such as swimming, snorkelling, boating and fishing could be expanded. Commercial possibilities would be created for hotels, tourist facilities and supporting services. The economic benefits would consist of a projected increase in producers' surplus, and possibly consumers' surplus. Indirect benefits would also be generated on domestic and international airlines owned by Indonesia as a result of an expanded tourist trade.

Benefits to agriculture would be generated in irrigation districts downstream from effluent discharge points and runoff areas as a consequence of reduced sediment and toxic substance loads. Dose-response functions might be used to estimate changes in physical yields for crops and other agricultural products. The increase in benefits would be the projected increase in producers' and consumers' surplus. Switching to new, more profitable crops might also be possible leading to further increases in economic welfare. Other benefits to agriculture would consist of reduced water treatment costs and lower operating costs for irrigation equipment such as pumps, sprays and piping.

The main benefits to industry would consist of reduced cost for the treatment of intake process water, cooling water and boiler feed. Such benefits can be estimated through cost-analysis techniques.

In upper catchments, a reduction in sediment loads and other pollutants would bestow a number of economic benefits such as reduced siltation of reservoirs. The benefit can be measured as the additional benefits accruing through an extension in the life of the reservoirs or as the postponement of the cost of constructing replacement reservoirs.

Further benefits would accrue to municipal water authorities through a reduction in water treatment costs, for example the cost of eliminating algal blooms in reservoirs, or the removal of organic waste, solids and pathogens before supplying water to urban consumers.

A reduction in sediment transport and deposition could result in a lower risk of flooding, especially in downstream areas. The economic benefit of reduced flooding should be measured in terms of a reduction in the expected damage costs. Flood events are specified in terms of the return period or probabilities of exceedance. The expected damage costs should be estimated by weighting the event-related damage costs by the corresponding annual probabilities of occurrence.

Sedimentation can also adversely affect water transport. To keep channels clear, both in rivers and in the bay, dredging is required. Any reduction in dredging operations as a result of sediment control would constitute a benefit of environmental improvement measured in terms of the reduced net cost of dredging operations.

To capture the full benefits of environmental improvement, it is essential that a long term planning horizon is selected. Further, the benefits of pollution control programmes should be assessed with respect to the projected situation without controls, rather than just the current situation. The present value of (future) benefits can be calculated using the prescribed social discount rate.

Economic Costs of Control

The direct costs of environmental improvement options can initially be measured as financial costs. Taxes, subsidies, and depreciation allowance all affect the ultimate cost impact. Economic measures of options, however, should be based on opportunity cost of inputs. Measurement of costs is usually more straightforward than measurement of the benefits, although difficulties may be created by non-marketed inputs. For example, labour required to carry out reafforestation projects in the catchment may be supplied by women who are unpaid. The value of such labour would be a legitimate cost component.

General cost categories cover capital costs and operating costs. Capital inputs enter the analysis at the time of commitment of resources. Operation and maintenance costs should be accounted for over the expected life of the pollution control equipment or up to a designated planning horizon. All cost estimates should be in real terms, free of future inflation.

For some improvement options, the opportunity cost may be measured in terms of defensive expenditures or avertive behaviour. For example, moving an industrial plant from one location to another to reduce the pollutant loadings on a particular stream may result in a loss of producers' surplus. Such a cost should be included in the cost of the control option. Relocation of households to reduce their exposure to environmental hazard may involve similar costs, including the costs of transporting possessions, legal fees, possible loss of capital and imputed costs associated with the stress of moving to an unfamiliar neighbourhood.

As in all economic analyses, the present value of costs should be calculated using the social discount rate.

Comparison of Benefits and Costs

For each pollution management option, the present value of net benefits (NPV) can be calculated by subtracting the present value of costs from the present value of benefits. Efficiency in the use of resources requires selection of the pollution control option that maximises NPV or, equivalently, minimises the sum of control costs and environmental damage costs.

Other considerations may, however, be important, such as the incidence of benefits and costs. At present, many of the environmental costs are being borne by subsistence fishing communities whose livelihoods are under threat from pollution and declining fish yields.

The incidence of improvement costs depends to a large extent on specific pollution control policies that are adopted. Under the "polluter pays principle" the costs of control must initially be borne by polluters themselves. In the Jakarta region a heavy cost burden would be incurred by manufacturing industries and by public authorities responsible for waste disposal. Such costs are often passed on to the general community by raising prices. The ultimate incidence of costs can be assessed only when such indirect effects have been taken into account.

Fiscal and Financial Impacts

Many of the benefits of improved water quality in the Jakarta region would consist of an increase in money income, especially in the tourism, fisheries and aquaculture industries. An increase in money incomes of the general population is also likely as a result of improved health and labour productivity. The net effect on manufacturing and industries is more difficult to assess, as these activities would incur both benefits and costs from water pollution control.

Increases in money incomes could provide a basis for public assistance or direct involvement in pollution control programmes. An increase in government taxation revenue would occur as a result of higher personal and corporate incomes. Special taxes could be applied to the sectors enjoying significant benefits, such as tourism. Entry fees to marine parks and other protected areas would comprise an additional source of government revenue.

An expansion of public revenue would permit the government to facilitate investment in public waste treatment facilities such as sewage treatment works and disposal systems for solid waste and trash. Special financial assistance could also be given to manufacturing industries facing economic difficulties in the treatment of waste. This would be especially important in the control of toxic substances such as heavy metals, PCBs and other carcinogens.

Use of Pollution Standards

If, for a variety of reasons, the benefits of pollution control options are too difficult to measure, the fallback is to use environmental quality standards in conjunction with cost-effectiveness analysis. Such standards may apply to discharges, ambient concentrations in the environment, exposure times of receptors or maximum doses of pollutants.

Judgements about standards are not easily made. The opinions of scientific experts such as health specialists, engineers or biologists, can be helpful. There is often a temptation to adopt the standards set by other countries or international authorities, but these may not be appropriate under local conditions. This is especially true of discharge standards, which have different effects on ambient environmental quality in different locations, depending on the pattern of discharges and natural system responses.

In setting standards, the cost of attainment may be an important decision variable, as the appropriate technology to yield a least cost solution may vary with the degree of stringency required.

To be effective, standards must be enforced. Studies show that in the Jakarta region, water quality standards established by the City Government of Jakarta have been frequently transgressed (Center for Environmental Studies, Gajah Mada University 1987; Soemantojo and Arief 1987). This has occurred for cadmium and chromium concentrations in river water, which are often five times the permissible level for drinking water, fisheries and agriculture. Lead concentrations in some rivers are up to twenty times the prescribed standard. In demersal fish, standards have been exceeded for cadmium, chromium, copper, mercury and lead. Standards have also been exceeded for organic waste and bacteria. Coliform counts in rivers are high. In Jakarta Bay counts of up to 10,000,000 per 100 ml have been recorded, and high concentrations of pathogens have been found in shellfish. Salmonella, Shigella, Eschericia and other pathogens known to cause stomach disease are present. BOD loads in rivers, associated with urban and human waste and with industrial discharges have been severe.

Institutional and Administrative Aspects

(Ben Boer and Angus Martyn)

Environmental and Administrative Policy in Indonesia

Indonesia has recognised its severe environmental problems since the early 1970s with the creation of a special unit on natural resources and environmental management within the national government. In 1978 a State Minister for Development, Supervision, and Environment was appointed. In 1983 this position changed to State Minister for Population and Environment. This Ministry is responsible for the coordination of environmental and natural resource management and its incorporation in development planning. It does not have any substantial decision making or coercive powers. The Ministry has a minor presence at the regional level, maintaining small offices in the provincial capitals.

The Environmental Impact Management Agency, or BAPEDAL, was created by Presidential Decree in 1990. BAPEDAL's mandate includes the development and implementation of policies on pollution control, hazardous waste management and environmental impact assessment. Although at present BAPEDAL is based in the capital, Jakarta, there are plans to establish regional offices.

In 1987/88 all government agencies were instructed to develop mechanisms for implementing environmental laws and regulations. Typically, however, fluctuations in Ministerial priorities and the significant time-lag between the adoption of policy and the effective operation of implementing mechanisms means that progress has been relatively slow in this area.

Environmental Legislation and Administrative Guidelines

In 1982 a statute titled *Basic Provisions for the Management of the Living Environment* was enacted. This statute provides the basis for environmental policy and legislation for the whole of Indonesia, including natural resources management, conservation, pollution control and environmental impact assessment. The Act recognises that the management of the living environment requires "an integrated effort in the utilisation , regulation, conservation, supervision, control, restoration and development of the living environment" (Article 2). It also recognises that the management of the living environment is "based upon the sustenance of the capability of the harmonious and balanced environment to support continued development for the improvement of human welfare" (Article 3).

In the "Elucidation" which accompanies this statute, it is stated that management of the living environment requires a system with integration as its main characteristic. The need for a national policy on management of the living environment has been recognised. It is also stated that a system for the protection and management of the living environment must be based on a clear, firm and comprehensive legal foundation. The Act thus serves as a basis for the evaluation and adjustment of all legislation containing provisions concerning aspects of the living environment, ie irrigation, mining, energy, forestry, protection and conservation of nature, industry, settlements, spatial organisations, land use, etc.

Environmental Impact Assessment

Article 16 of the *Basic Provisions* statute lays the basis for legislation concerning environmental impact assessment:

16. "Every plan which is considered likely to have a significant impact on the environment must be accompanied by an analysis of environmental impact, carried out according to government regulations."

This Article is explained in the "Elucidation":

> "In principle, development projects and activities produce impact on the living environment. Early planning of any development effort or activity must include a consideration of its major impact on the living environment, both physical and non-physical, including socio-cultural, so that an assessment can be made as to whether an environmental impact analysis should be carried out."

It is stated that such analysis should indicate more precisely the negative and positive impacts of a particular activity so that steps may be taken as early as possible to abate its negative impact and develop its positive impact. An attempt is made in the "Elucidation" to spell out some of the major impacts that should be considered; these include: the total number of people affected; the size of the area affected; the length of time during which the impact will persist; the intensity of the impact; the number of other environmental components affected; the cumulative nature of the impact; and whether the impact is irreversible.

Implementing regulations of EIA were passed by the national government in 1986 and became effective the following year. These regulations and a number of associated Ministerial guidelines provided the framework for comprehensive EIA in Indonesia. Both existing and proposed developments were to be subject to the EIA process. EIA can be carried out on either a central or provincial basis. At the central level, each government department or authority was directed to establish an EIA "Commission", charged with overseeing EIA of all activities connected with its portfolio responsibilities. The membership of these Commissions is laid down by Ministerial guidelines. Regional Commissions have also been established within the provincial governments, and these deal with developments not evaluated at the central level. Due principally to financial and human resources shortages, the application of the EIA process has not been consistent either at the central or regional levels. The difficulties encountered by the Indonesian government in introducing EIA are comprehensively discussed by Dick and Bailey (1992). One of the major problems is the lack of an effective administrative device linking EIA, development control and pollution licensing.

Pollution Law

The *Basic Provisions* statute includes the following Article on pollution control:

> 17. "Provisions on the overall and sectoral prevention and abatement of pollution of the living environment and its control shall be established by legislation."

To date this provision has not been followed with appropriate legislation.

Institutional Responsibilities in the Jakarta Region

Institutional and legislative aspects of pollution control in the Jakarta region have been described by Soetarjono (1987). Some of the main points are repeated here.

The City Government of Jakarta bears responsibility for management of the Jakarta region. The Governor has full executive powers. General development policies are formulated by the regional secretariat. The Regional Planning Board directly assists the Governor. The jurisdiction of the Regional Secretariat and the Regional Planning Board is confined to the City of Jakarta. Various decisions, announcements, instructions and regulations have been issued by the Governor to address problems of pollution.

Administrative tasks are carried out by the Regional Administration and by specific departments and Regional Enterprises. Implementation is the responsibility of the Inspectorate of the Provincial Region. Local Administration is guided by Mayors of Administrative Cities. Technical and service staff are located in technical departments.

Attempts by the City Government to address the pollution problem over the past two decades have been limited and largely ineffective. Regulation no 12 of 1971 required that waste-producing industries and/or activities were to be prohibited unless they installed approved pollution control equipment. Licensing requirements were introduced, applications for licenses were to be accompanied by a site plan, the pollution control method to be used, and the waste disposal process. The degree to which licensing requirements have been implemented is unknown.

Concerns over increased industrial development and expansion, and the corresponding threats to water, air and soil prompted a "Decision" of the Governor of Jakarta (no 484 of 1972). This set water quality criteria for rivers on the basis of their planned use. It stipulated that waste disposal into the rivers should not cause deterioration of the water quality below these criteria. However, guidelines were not given indicating how much waste could be disposed of without causing such deterioration.

The ineffectiveness of these measures led to Announcement of the Governor of Jakarta (no. DB. 12/1/4/1973) in 1973. This required industrial companies and other organisations, both public and private, to conform to the existing provisions regarding prevention of air, water and offshore pollution.

Governor of Jakarta Instruction, no. 203 was issued in 1978, apparently in response to an Act issued by the Minister of Health in the previous year regarding water quality standards. The Instruction points out that industrial enterprises of the Jakarta area have a duty to set up efforts to overcome the problems of environmental pollution. The Instruction is also directed to all supervisory departments in Jakarta. These departments were instructed to intensify supervision on industries, and to require owners of industries to install pollution control equipment.

In the Jakarta region, the natural boundaries are defined by catchments and coastal areas. Although the City of Jakarta has wide ranging powers, the responsibilities for various environmental management functions remain undefined (Soetarjono 1987). In addition, many important development activities, such as urbanisation and industrial development in the upper areas of the catchments, fall outside the jurisdiction of the City Government. Decisions taken in such areas can, nevertheless, affect environmental quality in the Jakarta region.

By the late 1980s it was recognised by the Indonesian government that the control of surface water pollution must be given the highest priority. Accordingly, in 1989 the PROKASIH, or clean waters programme, was introduced. This was a national initiative sponsored by BAPEDAL. Under this programme, a number of key river systems were targeted, including the Ciliwung, Cipinang and Mookervaart rivers, all of which flow through Jakarta. In essence, the PROKASIH strategy has been to negotiate individual written agreements with the management personnel of industries identified as point sources of excessive pollution. These agreements commit industries to progressively reduce emissions according to a set time-frame. Although there have been problems with enforcing agreements, the programme has generally resulted in environmental considerations being included in corporate policy decisions (Santosa 1991). At this stage, PROKASIH has concentrated on industrial sources of pollution. Over the medium term BAPEDAL has stated that PROKASIH will aim for a significant reduction in the amount of domestic waste entering river systems.

In 1990, Government Regulation no. 20, which gives responsibility for water pollution strategy to provincial governors, was promulgated. It appears, however, that this regulation has only partially been implemented.

As indicated by the case study, it is imperative that an integrated approach be taken to achieve an improvement in environmental quality in the Jakarta region.

References

Center for Environmental Studies, Gaja Mada University (1987) 'Mapping of Heavy Metals and Development of a Monitoring System for Jakarta Bay', Doc No 8/AWP/87,

Workshop Additional Working Papers, Workshop on Coastal Waters Management and the Bay of Jakarta, IFIAS - University of Indonesia, Jakarta.

Danusaputro, S.M. (1978) 'Environmental Legislation and Administration in Indonesia', Alumni, Bandung.

Dick, J. and L. Bailey (1992) *Indonesia's Environmental Assessment Process (AMDAL): Progress, Problems and a Suggested Blueprint for Improvement*, BAPEDAL, Jakarta.

Ongkosongo, O.S.R, S. Hadi, Y. Kartawiria, R. Ariani and S. Setiadi (1987) 'The Natural System and Environmental Quality Aspects of Jakarta Bay', Doc No 4/WP/87, Workshop Working Papers, Workshop on Coastal Waters Management and the Bay of Jakarta, IFIAS - University of Indonesia, Jakarta.

Risch, L. (1987) 'The Modelling of Aquaculture Water Demands and Production in the Brackish Water Areas of Jawa Barat', Memo 87/60.

Santosa, A. (1991) 'Prokasih: A Brief Analysis', unpublished paper.

Soemantojo, R. and A. Arief (1987) 'The Source of Pollutants in Jakarta Bay', Doc No 4/SI/87, Workshop Supporting Information, Workshop on Coastal Waters Management and the Bay of Jakarta, IFIAS - University of Indonesia, Jakarta.

Soetarjono, R. (1987) 'Legal and Institutional Setting of the Jakarta Bay Management', Doc No 9/WP/87, Workshop Working Papers, Workshop on Coastal Waters Management and the Bay of Jakarta, IFIAS - University of Indonesia, Jakarta.

Suprianto, B., A. Arief, H. Syarain and N.E. Wahyuningsih (1987a) 'Waste Disposal in Jakarta Bay', Doc No 3/SI/87, Workshop on Coastal Waters Management and the Bay of Jakarta, IFIAS - University of Indonesia, Jakarta.

Suprianto, B., H. Syarin and S. Sotiadi (1987b) 'Marine National Park', Doc No 2/SI/87, Centre for Research of Human Resources and the Environment, University of Indonesia.

Swartz, A. (1992) 'Biting the Bullet', *Far Eastern Economic Review*, 23 July.

University of Indonesia Study Team (1987) 'Jakarta Bay Management Study, Phase I - Draft Final Report', Workshop on Coastal Waters Management and the Bay of Jakarta, IFIAS-University of Indonesia, Jakarta.

Wahyuningsih, N.E. (1987) 'Fishing in Jakarta Bay', Doc No 5/SI/87, Workshop Supporting Information, Workshop on Coastal Waters Management and the Bay of Jakarta, IFIAS - University of Indonesia, Jakarta.

Chapter 6

CONSERVATION FARMING IN NORTHERN THAILAND

TOR HUNDLOE[1]

This case study illustrates the application of economic evaluation methods to an important environmental problem in a developing country. Indeed, the problem consists of two interrelated problems. First, it concerns the deteriorating financial and social situation of hill tribe farmers in the Northern Thailand highlands, caused largely by increasing population pressure and the consequent need to produce more food for self-sufficiency from a finite and ecologically sensitive resource. Second, the problem involves adverse off-site environmental effects on other productive activities, in particular forestry operations and downstream uses of water.

On-farm soil erosion and degradation, the causes of decreasing yields and financial and social problems, have induced farmers to encroach upon the diminishing forests of Northern Thailand, thereby reducing the value of this productive natural resource. Soil erosion is also leading to siltation and sedimentation of downstream water storages, adversely affecting irrigated

[1] The author wishes to acknowledge the assistance provided by the project managers, Australian Agricultural Consulting and Management Company Pty Ltd, and in particular Mark Hoey. The contribution of Ray Moyle, who assisted with data collection in Northern Thailand, is also appreciated. John Dixon and David James read earlier drafts and made helpful comments. Any errors or omissions remain the responsibility of the author. Institutional support provided by the Environment and Policy Institute, East-West Center, Hawaii, during completion of the manuscript is also gratefully acknowledged.

D. James, The Application of Economic Techniques in Environmental Impact Assessment, 143–181.

cultivation and hydro-electricity generation, and contributing to sedimentation and flooding probably as far downstream as Bangkok.

The project discussed in this case study is a conservation farming project in the highland area of Northern Thailand, receiving funding assistance from the Australian International Development Assistance Bureau and managed by the Australian Agricultural Consulting and Management Company. It is known as the Highland Agricultural and Social Development (HASD) project.

The case study begins with a general description of natural resources and their uses in Thailand. Problems created by misuse of resources are then discussed, in particular the cultivation of land by hill tribes in Northern Thailand. An introduction is given to the political and institutional background with respect to environmental impact assessment and land-use policy. The HASD project is then described. An economic evaluation of the project is carried out, comparing the with- and without- project situations. On-site impacts are dealt with first, then off-site impacts. The study concludes with a comparison of benefits and costs, and draws out the main policy implications.

Natural Resources of Thailand

General

Thailand is a fertile country, known in earlier times as the "Golden Land" because of the high productivity of its farmlands, forests and the richness of its mineral resources. A thirteenth century stone pillar inscription (dating from the Sukhothai period of King Ramkhamhaing the Great) proclaims:

> "This Sukhothai is good. In the water there are fish, in the field there is rice. The King does not levy a tax on its people... Whoever wants to trade in elephants can do so. Whoever wants to trade in horses can do so. Whoever wants to trade in silver or gold can do so; people are contented..." (Arbhabhirama et al. 1987)

Thailand has an area of some 513,000 sq km. The country stretches 1,620 km from north to south and, at its widest, 780 km from east to west. It is bounded by Malaysia in the peninsular south, Burma to the west and north-west, Laos to the northeast and Kampuchea to the south east. Much of the country is mountainous, especially in the north. Some 52% of its area is cultivated and 30% is covered by forest, although this percentage is rapidly dwindling.

Thailand is divided into four regions:

- the central region with a fertile alluvial flood plain known as the Chao-Phraya River basin, which has traditionally exported rice to other parts of South-East Asia;
- the north-east, a semi-arid rolling plateau, once the centre of Thai agriculture;
- the north region, which is the most highly forested area, and includes intensive cultivation of temperate fruits and crops;
- the south, which supports rubber plantations and tropical fruit orchards.

A map of Northern Thailand, which includes the study area, is shown in Figure 6.1, with its provinces, major cities and water bodies.

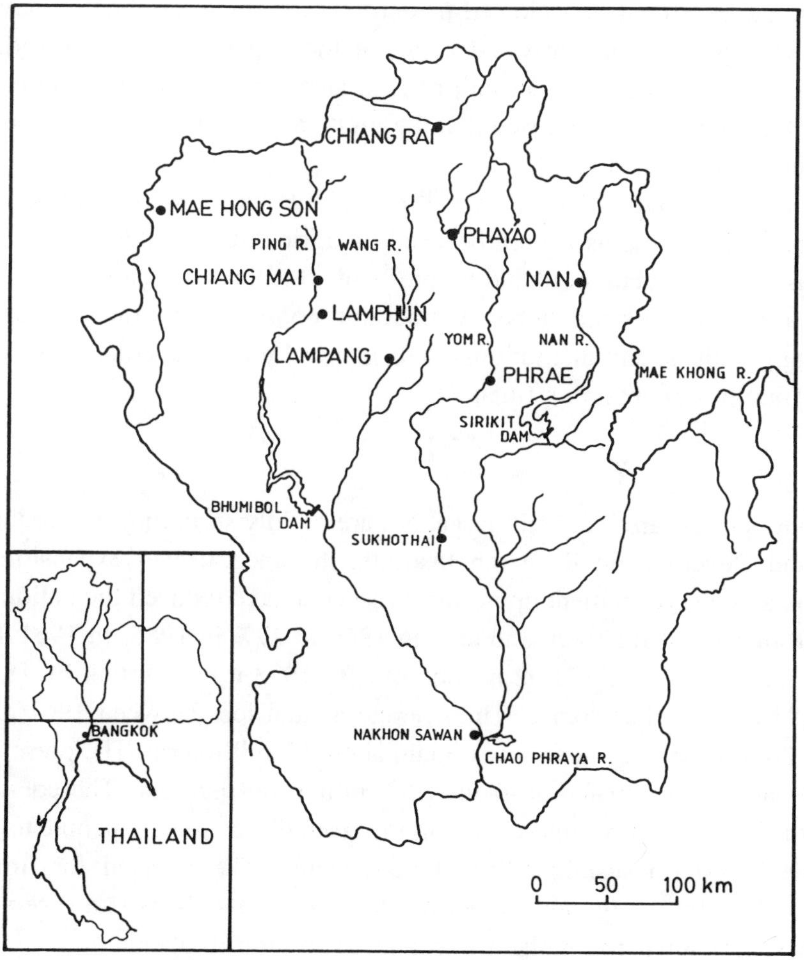

Figure 6.1. Map of northern Thailand.

Agriculture

Thailand is one of Asia's leading exporters of agricultural products. The major export crops are rice, rubber, cassava, maize and sugar. Approximately 80% of Thailand's population lives in the rural areas, earning a livelihood principally from agriculture. Agriculture contributed about 70% of gross domestic product in 1985, 60% of exports, and provided full or part-time employment for more than 70% of the nation's labour force.

Of the 152 million rai (25 million ha) of agricultural land only 21 million rai (14%) are irrigated. The remainder are rainfed. (1 rai = 0.16 hectares)

From 1960 to 1980, the value of agricultural production grew in real terms, but the rate of growth has slowed recently. Increases in production have come from the expansion of cultivated areas, at the expense of forests, rather than from increases in the yield per unit of land area. As Arbhabhirama et al. (1987) state: "It is generally recognised that the limits of the nation's land resources are now being approached."

Crop yields have fallen as a result of degradation of soils and poor farming practices. The problems generally faced by agriculture include: irregular rainy seasons; soil degradation, including physical, chemical and biological deterioration; cultivation on steep slopes without soil conservation practices; unsuitable farming practices; and institutional and financial problems (such as lack of title and difficulties in obtaining finance).

Forests

Thailand's forest areas are still large, but are rapidly shrinking. According to the Thailand Development Research Institute, the uncontrolled expansion of cultivation and encroachment upon forest reserves have reduced the nation's forest area from 50% of the total land area in 1960 to 29% in 1985. In 1961 Thailand had a forested area of 171.02 million rai. By 1973 it had shrunk by 18.9% and by 1985 by a further 26.5%. The average annual loss between 1961 and 1985 was 3.24 million rai. There now remain about 93 million rai. The present rate of deforestation is estimated at about 1.52 million rai per year. The country's forests fulfil a number of functions, ranging from direct uses by people to supporting the long-term sustainability of ecosystems. The demand for timber and fuelwood is increasing, and timber has to be imported. This is a reversal of the earlier situation where Thailand was a net exporter of timber.

In Northern Thailand, shifting cultivation is resulting in significant forest destruction. The loss of forest in itself is important. With encroachment on steep land (slope greater than 35%) the consequent soil erosion represents a further diminution of the nation's natural resources. Soils which support virgin tropical forests are characterised by low soil bulk density and high macro-porosity, partly due to the relatively high activity of soil fauna. Once the forest cover is removed, the soil characteristics begin to change. These changes have a significant influence on the erodibility of the soils.

Water

Thailand receives approximately 800 billion cu m of rainfall per year. While this may sugggest that water is plentiful, the fact is that 85% cent of the rainfall is received between May and October. To meet increasing demands for water, development of water resources is essential.

It is believed that Thailand's first reservoir was built about 700 years ago. Irrigation has since been an essential component of the country's economy. Nevertheless, there are significant water shortages in some regions; or there are floods. In the central region many areas are inundated in the rainy season, but in the dry season water has to be imported from the north. Only 30% of the irrigable area covered by the Greater Chao Phraya Project has adequate irrigation in the dry season (Arbhabhirama et al. 1987). Demand for water for all uses (agricultural, domestic, industrial and waterway transport) is continually growing, and competition exists between users. Most obvious reservoir sites have already been used. Water resources are subject to decisions of numerous government agencies with overlapping and, in some cases, competing objectives.

Dams

The first multi-purpose dam with large-scale hydro-power constructed in Thailand was the Bhumibol Dam on the Ping River in Tak province. This was followed by the Sirikit Dam on the Nan River. Both dams are located in Northern Thailand (see Figure 6.1). There are numerous smaller reservoirs in the catchment areas.

The sharp increase in demand for electricity following the Second World War led to construction of these multi-purpose dams. Hydro-power accounts for about 22% of the nation's energy requirements. Hydro-power costs 0.50 Baht/kWh, while thermal power costs 1.10 Baht/kWh, measured in 1985 prices.

The other major purpose of these dams is irrigation. The irrigation benefits derived from the Bhumibol and Sirikit dams include: (i) dry season rice production of approximately 2.5 million rai; and (ii) an increase of about 30% in wet season rice yield due to more reliable water supply (Social Research Institute, Chulalongkorn University 1987). The dry season production is expected to increase over time. Additional benefits of the dams include fish production, flood control and prevention of salt water intrusion in the Bangkok area.

The Bhumibol dam is a large storage reservoir with a total surface area of 318 sq km and a total volume of 13,462 million cu m at maximum storage level. The impounded reservoir is approximately 207 km long with a maximum width of 7.5 km. At present, seven turbine and generator units have been installed and are in operation, with a total capacity of 535 MW. Electricity generated by the Bhumibol dam is transmitted to a substation at the damsite and other substations in Nakhon Sawan, Ang Thong and Bangkok. Water released from the Bhumibol dam is used to irrigate a number of paddy fields in several provinces in the Chao Phraya River basin.

The Sirikit dam is a multi-purpose dam included in the Nan River Basin Development Project. The other two diversion dams are the Uttaradit and Naresuan dams. The Sirikit dam was constructed on the Nan River in Tha Pla district, Uttaradit province, about 58 km east of the city of Uttaradit. It is the largest earth-filled dam in Thailand. Its construction resulted in a large storage reservoir with a total surface area of 260 sq km and a storage capacity of 10,500 million cu m at maximum storage level. The powerhouse of the Sirikit dam is equipped with four turbine and generator units with a total capacity of 500 MW. Apart from electricity generation, the Sirikit dam, together with the Naresuan dam in Phitsanulok, can supply water to a number of irrigable lands in the Nan and Chao Phraya river basins.

The total inflows (that is, total outflows plus changes in storage) of the Bhumibol and Sirikit dams between 1975 and 1984 were "roughly constant" (Arbhabhirama et al. 1987).

In recent years, the area of irrigation has increased throughout the country, and in every region. The irrigated area in the Northern region is increasing at a greater rate than in the Central region. Arbhabhirama et al. (1987) explain:

> "Assuming this trend continues, a point will be reached in the near future where water released from the Northern region, after satisfying its own requirements, will be less than at present. As a consequence, the demand for irrigation water in the Central region will not be fulfilled."

Soil erosion is a problem. According to the Electricity Generating Authority of Thailand (in a report prepared by the Social Research Institute, Chulalongkorn University):

> "Forest encroachment in the watershed areas of the Bhumibol and Sirikit dams will have long-term effects on reservoir operation, since it will increase the rate of soil erosion. It has been estimated that the shift from forest land to agricultural land would increase the rate of soil erosion about 6-10 times, thus reducing the useful life of the impounded reservoirs."

Use of Resources

Trends in Population

In the past century, population growth (and the consequent need to produce food, shelter and clothing) put immense pressure on the natural environment. In 1911 the population was 8 million. By 1984 it had passed 50 million, with a growth rate of 2% per annum. It is predicted to be between 64 and 68 million by the year 2000. About 80% of the population lives in rural areas, although the highest concentration is in the Bangkok area. There is a continual drift to the cities.

Hill Tribes

The highlands of Northern Thailand have been inhabited and cultivated by swidden (slash and burn) agriculture since the first hill tribe people entered the area several centuries ago. The term "hill tribe" is a generic name for tribal groups who over the centuries have moved from China into the mountainous areas of Burma, Laos and Thailand.

The major immigration into northern Thailand occurred in the last few decades. The hill tribe population is scattered over 21 provinces of Thailand and the total population is estimated at 500,000 persons. Chaturabhad (1988) describes 10 tribes: Karen, Meo, Lahu, Yao, H'tin, Akha, Lisu, Lua, Khamu and Mrabri. The tribes which traditionally grew opium are the Meo, Yao, Lahu, Lisu and Akha. Those which are not traditional opium growing tribes are the Karen, Lua, H'tin and Khamu. The Mrabri are traditional hunters and gatherers.

Cultivation Practices of Hill Tribes

Ethnographic studies of the hill tribes have been undertaken by Walker (1986) and Chaturabhad (1988). While there are significant cultural differences among the tribes, some generalisations can be made. Until recently, the economy was a subsistence one, based on rice grown on steep slopes. The cash crop for which the hill tribes are best known is opium. In recent times, as a result of gradual integration into the cash economy, a variety of other crops such as maize, vegetables and fruits, have become important components of the hill tribe economy.

Swidden agriculture has led to the destruction of a significant area of forest in northern Thailand. Although having relatively benign environmental impacts with small populations, swidden results in major on-site and off-site effects as populations grow. Land and soil are finite resources and in heavy rainfall areas, such as in Northern Thailand, soil erosion and degradation are inevitable effects of overcultivation. Population pressure hastens erosion and degradation. Population growth has led to increasing destruction of virgin forest to make way for swidden cultivation.

Laws have now been proclaimed in Thailand aimed at protecting the remaining forests. This has imposed a restriction on the ability of the hill tribes to expand the area under swidden. The consequence has been a reduction in the swidden cycle and the acceleration of soil erosion and degradation. As land is left in fallow for shorter periods, the replenishment of soil nutrients and chemicals has diminished, leading to reduced crop yields. Lower yields reduce the subsistence production of cultivators and the value of cash crops if they are grown.

Major problems facing the hill tribes are: (i) poverty, with over 70% of the hill population not self-sufficient in rice; (ii) soil erosion, with losses of 50 to 100 t/ha/y (considered to be 10 to 20 times the acceptable limit); (iii) deforestation, with approximately 2% of forest area being lost annually; (iv) social problems, with the hill tribes requiring health and education facilities, citizenship rights and land tenure; (v) road access and other infrastructure, which are presently inadequate; (vi) security from incursions from Burma and Laos; and (vii) smoking and trading of opium.

The off-site consequence of these economic and social pressures are also important. The eroded soil leads to siltation and sedimentation of the streams and, ultimately, the major river systems into which they feed. The result is siltation in dams, potential loss of water for irrigation and hydro-power purposes, and increased probability of downstream flooding.

These on-site and off-site problems have recently been addressed through the implementation of various land use and social development projects such as the HASD project.

Environmental Problems

There are signs that the present pattern of development may be unsustainable. In many regions of the country high per unit yields of paddy rice can be produced - over one tonne per rai in several irrigated areas. In areas of traditional rainfed rice production, output averages about one-fifth that of irrigated areas. The fish catch per capita, from both inland and coastal waters, is falling. The nation's water resources are under considerable strain from domestic use in Bangkok and other cities, and from irrigation. Water conservation measures are becoming increasingly necessary to ensure a balance between demand and supply in terms of both quantity and quality.

Soil Erosion

Soil erosion is one of the most pressing natural resource problems in Thailand. Srikhajon et al. (1980) estimated that for Thailand as a whole approximately 107 million rai suffer from medium or high degrees of soil erosion, particularly upland areas where the slopes are greater than 5%. Arbhabhirama et al. (1987) report various findings. In Nan province, the average annual soil loss in many areas exceeded 16 tons per rai - 20 times the recommended tolerable annual loss. In Phatchabun province, in an area with a 9% slope, soil losses in 1983 were 162.6 t/ha/y. In Chonburi province, losses were 110 t/ha/y from lands with 3% slopes and bare soil. In Kalasin province, on 6% slopes of bare soil, a loss of 20.8 t/ha/y was recorded.

A comparative analysis of soil loss on various land management plots has shown that soil loss from forest management is far less than from swidden and continuous rice cultivation (Arbhabhirama et al. 1987). The effect of producing one crop of rice every 10 years is to increase soil loss by 1.5 to 1.7 times the losses in a typical forest. If the swidden cycle is shortened from once in ten years to once in two years there is a 1.6 to 1.8 times increase in soil loss. These figures suggest that soil loss from a short swidden rotation is 2.4 to 3 times greater than from forest management. Arbhabhirama et al. contend that if land uses change from forest to row crop cultivation, without soil and water conservation measures, the rate of soil loss will increase about 6 to 10 times.

Soil erosion and degradation, if unchecked, will inevitably result in declining yields. The following data illustrate the point. Corn yields on newly cleared areas average 262 kg/rai whereas those in old areas average 166 kg/rai; upland rice yields average 221 kg/rai on new areas but only 114 kg/rai on old areas (Arbhabhirama et al. 1987). Soil erosion experiments undertaken for the HASD project, discussed later, have given similar results. A serious consequence of decreasing yields is the inducement for encroachment upon forest lands, to replace degraded land returned to bush fallow.

Arbhabhirama et al. (1987) have assessed the soil conservation situation as follows:

> "Soil degradation problems, including declining soil fertility, soil compaction, soil erosion and salinization, threaten to become more prominent. More attention must be given to soil conservation since past programs have rarely been successful.
>
> "The limited success of past soil and water conservation projects has been due to three factors. There has been a lack of support from the government and the responsible organisations were not prepared to tackle the problems effectively. Another negative factor has been the social and economic conditions of the farmers. Thirdly, the conservation techniques were difficult to use and no immediate return could be expected.
>
> "Soil and water conservation projects should have the following characteristics: (i) they should be simple enough for individual farmers to apply; (ii) the cost should be low; (iii) the project should be farm-based; (iv) less construction should be required; and (v) the project should be easily replicable."
>
> "These, though they seem rather simple, require some changes in the current programs. Moreover, there is also a need for new and simple techniques of conservation which must be adopted and tested carefully before introducing to the farmers."

The conservation farming system concept developed by the HASD project is consistent with the approach suggested above.

Flooding

According to the Social Science Research Institute (1987), although the flood control benefits of the Bhumibol and Sirikit dams are "still in question", the dams have played...

> "a noticeable role in flood protection at the downstream flood prone area. They have provided a reduction of flood peak, flood volume and shortened flood durations. These have resulted in the reduction of agricultural damages markedly. However, it is realized that the Bhumibol and Sirikit reservoirs cannot completely protect flooding in the lower Chao Phraya basin. This is because of the amount of water that is uncontrolled and is still plentiful during the wet season. Thus, flooding the lower Chao Phraya area cannot be avoided if other measures do not comply".

The city of Bangkok is located on alluvial soils near the mouth of the Chao Phraya River. It is a city of some 6 million people. The city is frequently flooded, for two major reasons. First, the city is "sinking"; that is, land subsidence is occurring because of over-pumping of ground water. The second reason is the heavy monsoonal rainfall which overtaxes the capacity of the river. The situation is aggravated by the fact that the city's drainage system has not kept pace with rapid development.

A flood in 1982 caused damage valued at 6,600 million Baht. Without the major dams upstream the flooding problem could have been much worse. Any loss of the capacity to hold water, by sedimentation, could exacerbate the flooding problem. This again points to the potential benefits of controlling soil erosion in the catchment area.

Water Quality and Fisheries Production in Reservoirs

As of 1987, the water quality in the Bhumibol dam was considered suitable for all aquatic life (Social Research Institute 1987). The dam had 34 species of fish, belonging to 15 families, with carp being dominant. There were 475 fishing households, most of them in the upper section of the reservoir, earning an average daily income per fisherman of 39 Baht. About 800 tons of fish are caught annually.

Water quality in the Sirikit dam was described by the Social Science Research Institute (1987) as "rather good". There were 31 species of fish in 18 families, with carp being the dominant species. Around 550 households engaged in fishing, with an average daily income per fisherman of 60 Baht. There was some aquaculture in the reservoir, but its extent is not documented. The annual catch from the reservoir is around 1,000 tons.

Siltation and Sedimentation

The streams in Northern Thailand carry high sediment loads as a consequence of soil erosion. The sediment load into Bhumibol dam has been estimated to reduce its capacity by 9.3 million cu m of storage annually, equivalent to 0.07% of the total storage. The annual reduction of capacity of the Sirikit dam has been estimated at 6.0 million cu m annually, or 0.06% of its total storage (Attaviroj 1986). A number of smaller reservoirs in northern Thailand are similarly affected. Reduced sedimentation has occurred at the mouth of the Chao Phraya River as a result of the dams and sedimentation of water courses.

Attaviroj (1986) has estimated the annual storage depletion of the major dams and reservoirs in Northern Thailand as 18.18 million cu m. This storage depletion represents a loss of 1,164 ha (7,272 rai) of irrigable land per annum. The forgone net annual income from rice production is 4 million Baht. The relative contribution of upland cultivation to soil erosion and subsequent income losses is not easily determined, although rough estimates can be made.

Political and Institutional Setting

(Ben Boer and Angus Martyn)

Political Setting

Thailand has a long history of human settlement, comprising many small kingdoms and city states which over time amalgamated to form the Kingdom of Siam. The modern Kingdom of Thailand dates from 1939. Under the constitution of 1978, politically the country was a constitutional monarchy. However, following a period of political instability in 1990-91, this constitution was abrogated and an interim constitution was put into place in 1991, which is still in force. Whilst the King remains head of state, the exact relationship between the other arms of government - the legislature, the executive and the judiciary - is uncertain.

Development Planning

The National Economic and Social Development Board is the government's main planning body. It prepares national level and macro-economic plans. It is also primarily responsible for preparing, monitoring and evaluating the 5-year national economic and social development plans. It is the central point in the planning system.

The Thai Government began a series of 5-year national socio-economic development plans in 1961 with the formal incorporation of environmental matters into the national planning process in the fourth national plan of 1977-81. The seventh national plan (1992-1996) focuses on five major areas of environmental management policy:

1) Natural resource management.
2) Environmental quality.
3) Energy and environment.
4) Industry and environment.
5) Urbanisation and environment.

The plan recognised that the system for the administration of natural resources was in need of improvement. To this end a number strategies were proposed, including the facilitation of public participation in conservation projects, the reduction of conflicts, by the use of land use plans, in relation to utilising natural resources, the imposition of monetary and fiscal measures so that the public receive socially just benefits from the use of natural resources; tying the planning process more closely with the need to protect systematically vital areas, forests and coastal areas. Under measures to improve the quality of the Thai environment, the plan has the objective of the protection of one quarter of the total land area under the designation of "conservation" forest.

Policies on the Environment

The 1978 Constitution incorporated two sections on environmental matters as follows:

1) Section 65.

> "The State shall preserve, maintain an environmental balance and shall eliminate pollutions which endanger health and hygiene of the population."

2) Section 69.

> "The State should promote the exploration of natural resources in order that they may be economically exploited for the Thai people, provided that it is not contrary to the principles of conservation."

Subsequently, a national environmental policy was formulated and published as "National Policies and Measures on Environmental Development". The main objectives of this policy were:

1) To promote protection, rehabilitation and conservation of natural environmental resources on a continuing basis for supporting development activities and quality of life so they will be available for posterity as well as now.
2) To maintain desirable minimum standards of environmental quality for the population, particularly for reasons of health, hygiene, aesthetics and orderliness, these being the main ingredients enhancing the quality of life.
3) To ensure appropriate utilisation of natural resources so as to yield maximum social benefits at minimum cost to national environment while aiming to prevent any danger to public health and safety as well as enabling the recycling of the natural resources used.
4) To create a balance between the population levels and available natural resources in order to maintain a proper quality of life for the population.

5) To create a balance between development and conservation of natural resources in the development of urban areas in order to upgrade the standard of living.
6) To preserve natural scenic values and sites of buildings having special significance from the natural, historical, archaeological, cultural and religious points of view.

In setting out this policy it was recognised that the environmental problem was very urgent, that the natural environment and resources are part of the national heritage, that an essential component in overall environmental protection was the preparation of EIA reports for both the private and public sectors, the setting of appropriate environmental standards, adequate budgetary support, provision for study and research on important environmental issues and environmental education to raise public awareness (Panat 1983).

This policy is likely to be revised if, as anticipated, a "national environmental master plan" similar to the National five year plans eventuates (Board of Investment 1992).

Environmental Legislation and Administrative Structures

Environmental problems have accompanied Thailand's general social and industrial development. Since the 1970s the Thai Government has instituted national measures in order to meet the problems of environmental degradation and pollution. The National Environment Board (NEB) was established under the *Improvement and Conservation of National Environmental Quality Act* (NEQA) 1975 (amended in 1978). As an advisory board under the Ministry of Science, Technology and Energy (MOSTE), the NEB's powers did not include any decision making functions; these were retained within the Ministry itself. The Act introduced the concept of environmental quality standards, a requirement for environmental impact assessment and the use of "emergency powers". The Act also established two sets of policy guidelines on water pollution, for inland and marine waters.

It was obvious that the NEQA and its associated institutional agencies had become inadequate to the task of implementing a system of integrated environmental management. It was also considered inadequate in terms of implementation due to the ambiguity of some of its provisions (Kritiporn et al. 1990). Due to the large number of government agencies involved, the success of management programmes depend on effective coordination at all levels of planning and implementation. This coordination, it appears, was often lacking.

In the area of land-based resources, the situation seems to have been no different, with two dozen agencies under several Ministeries having responsibilities in the area and an antiquated land-law system. Arbhabhirama et al. (1987) comment:

> "Confusion, not surprisingly is the order of the day. This leads to delays and to inequities, particularly in terms of the widely different land title documents that participants in different government-backed programs receive from the respective agencies."

A radical overhaul of environmental legislation and institutions was completed in 1992. Probably the most important aspect was the replacement of the *Improvement and Conservation of National Environmental Quality Act* 1975 by a new Act of the same name. The 1992 legislation elevated the NEB to sub-cabinet level and gave it decision making powers that had previously been delegated to the Ministry of Science, Technology and Environment (MOSTE).

The National Environment Board

The "new" NEB consists of some 20 members. Eleven are Ministers of State, including the Prime Minister, who acts as chairperson. The remaining positions are occupied by several senior bureaucrats, including the Permanent Secretary of MOSTE, and six to eight persons outside the government sector. It is supported by a policy secretariat divided into three newly created "departments". These are:

1) The Office of Environmental Policy and Planning, which is charged with the development of Thailand's environmental policies, including the "environmental master plan" referred to earlier. Its mandate also extends to the supervision of provincial administrative agencies involved in environmental affairs.
2) The Pollution Control Department has the responsibility for undertaking comprehensive pollution monitoring and adopting appropriate responses. Current priorities include monitoring of air pollution in the Bangkok region, developing appropriate ambient air quality standards and monitoring heavy material concentrations and effluent levels in Thailand's rivers and klongs.
3) The Department of Environmental Quality Promotion: the duties of this department include information management and coordination, private and public sector liaison, promotion of environmental projects and providing advice on the utilisation of the 5 billion Baht (USD 200 million) Environment Fund. The fund will be used principally to finance pollution control by the government and the private sector.

Environmental Impact Assessment

EIA was first introduced in legislative form under an amendment to the old NEQA in 1978. This gave the NEB the authority to require EIA from project proponents in both the public and private sectors. This authority did not, however, extend to a direct power to monitor and enforce EIA recommendations. The NEB therefore had difficulty in preventing the implementation and operation of developments that failed to comply with the standards set under the EIA process (Kritiporn et al. 1990).

The *National Environmental Quality Act* 1992 provides for EIA. The first paragraph of s. 46 states:

> "For the purposes of environmental quality promotion and control, the [environment] Minister shall, with the approval of the National Environment Board, have the power to specify...types and sizes of projects and activities likely to have an environmental impact, or any government agency, state enterprise or private person, which are required to prepare reports on environmental assessment or seek approval in accordance with section 47, section 48 and section 49."

Where EIA is required, information is to be supplied in the form of an initial environmental examination (IEE). This is to be submitted to the Office of Environmental Policy and Planning. The Office may approve the IEE, or reject it on the grounds of incomplete data or documentation. If approved, the Office makes "preliminary comments" on the IEE and then refers it to a "committee of experts". This committee has the power to approve or reject the project or activity itself. If it fails to reach a decision within 45 days the assessment report is deemed to have been approved. If the committee approves the project, the authority which is legally empowered to grant permission to it must then give that permission, incorporating all conditions laid down by the committee. These conditions are legally enforceable under the appropriate laws. In order to maintain quality control of EIA, the Environment Minister may require any EIA report to be certified by a specially licensed EIA specialist.

The Office of Environmental Policy and Planning does not appear to have substantially revised the NEB manual of guidelines for the preparation of environmental impact evaluation published in 1979. The format of the EIA report required under the guidelines includes an introduction, description of the project, description of the relevant environment, anticipated environmental impact and plans for protection, the consideration of alternatives, arrangements for monitoring and summary conclusions. In the consideration of alternatives, where the proposed project will result in serious loss of environmental resources, the report must include consideration of alternative projects or approaches which could achieve the same or equivalent results (Wanichsombat 1988).

Land-Use Policy and Administration

In the recent past when population pressures were less significant, there was adequate land for every purpose - agriculture, human settlements and water bodies. Forest land was considered abundant. It was generally believed that forest land could be cleared whenever it was needed for other uses. With a rapidly increasing population and diminishing arable land per capita, widespread misuse of land has occurred, leading to cleared and abandoned land, under-utilized land and unnecessary fallow land.

Land tenure and land rights are complex and unresolved issues in Thailand. In 1985, only 12% of the 152 million rai of agricultural land carried title deeds, 48% was covered by a "certificate of utilisation" or other legal document and 40% had no legal entitlements attached. The *National Forest Reserve Act* of 1964, which prohibited the encroachment of national forest reserves, seems to have been largely ineffective. An immense area of illegally occupied land (55 million rai) has been illegally carved out of forest reserves or areas sought for forest reserves. Of this area, approximately 17 million rai has been degazetted as forest and has been allocated to various institutions, including the Agriculture Land Reform Office. In addition, vast areas of undeclared natural areas of forest have been occupied by physical encroachment. It appears that this type of encroachment has been twice as high as encroachment of legally declared forest reserves (Panayotou and Parasuk 1990). The existence of a multitude of landless and low-income farmers means that this is an on-going problem.

It would appear that improved land use planning and conservation farming systems must be introduced to obtain a sustainable maximum return from the already limited supply of land. A coherent land-use policy does not presently exist, although guidelines for the administration of land have been formulated by the numerous agencies with interests or responsibilities in land. The various agencies and their programmes often have widely different objectives. For example, agencies with charters aimed to enhance agriculture can differ in their objectives from those with charters to deal with forestry, water resources, the welfare of hill tribes, and so on.

The Thai government has in recent years been attempting to address land-use issues through various programmes of land allocation, land conservation, land reform and land use planning, and accelerating the granting of the titles to land.

The HASD Project

Physical Setting of the Project

Northern Thailand consists of 17 provinces and lies between latitudes 15° N and 21.5° N and longitudes 97.3° E and 102° E. It covers an area of approximately 106,027,680 rai. Northern Thailand can be divided by topography into two subregions. The upper region - the highlands - consists of eight provinces, Mae Hong Son, Chiang Rai, Chiang Mai, Nan, Lampang, Lamphun, Phayao and Phrae (see Figure 6.1).

The northern ranges and valleys occupy the area above 18° N latitude. There are five main parallel ranges running north-south. The ranges and the valleys between them provide the catchment and headwaters of the four major tributaries of the Chao Phraya River - the Ping, Wang, Yom and Nan Rivers which drain southward. Above the Phipannam Divide are basins which drain northward into the Mae Khong River, through the Mae Kok and Ing Rivers. In the northwest corner are watersheds which feed into the Salween River of Burma. The Chao Phraya basin is the most important and largest river basin in Thailand.

The Dan Laos mountain range, in the extreme north of Thailand, is the watershed between the Salween Valley in Burma and the Mae Khong in Laos. On the western side is the Tanon Tong Chai range, extending to the southwest along the Burmese border to join the Tenessarim range, which runs down to the Peninsula. The south is bounded by the extensive alluvial flat of the Chao Phraya River. The east is bounded by the Luang Phra Bang and Petchabun ranges.

The series of hill and mountain ranges, running approximately north-south, form the catchment areas for the Ping, Wang, Yom and Nan Rivers. These rivers join above the town of Paknampho to form the Chao Phraya River. The western mountain range is the headwater of the smaller rivers - the Mae Lamao, Mae Pai and Mae Yuan. These empty into the Salween River. In the northeast of the region, beyond the watersheds of the four main tributaries of the Chao Phraya River, there are a number of swampy basins. The trunk stream of these drains into the Mae Ink and Mae Kok which are both tributaries of the Mae Khong River.

The Geographic Setting

Northern Thailand can be divided into three major land forms: lowlands, uplands and highlands. The highlands can be defined as areas that are higher than 500 m. (Other definitions exist, for example above 300m or 400 m.) The range of altitudes is 500 to 2500 m.

The HASD project is confined to the highlands of Northern Thailand. The total land in the eight provinces in the HASD area is 8,585,200 ha (53,657,662 rai). The amount of farmland as of 1983 was 1,064,200 ha (6,651,190 rai). The highland area of the eight provinces (500 m and above) is 6,177,270 ha (38,607,939 rai) or 72% of the total land in the area. It is this highland area which defines the boundaries of the HASD project.

In terms of agriculture, a considerable amount of this highland area consists of paddy in the lower valleys, for example, Chiang Mai province. It is estimated that only 400,000 to 700,000 ha (2,500,000 to 4,375,000 rai) is in the swidden cycle. This amounts to one-third to one-quarter of the area. The remaining two-thirds to three-quarters is fallow. This suggests that, for the lower estimate, 100,000 ha are in productive use and for the higher estimate 230,000 ha.

Soil Erosion and Degradation

Soil erosion and degradation are major on-site and off-site problems in Northern Thailand. The soils vary with topography, from gravelly red-yellow podzolics and loamy red latosols on the high terraces to loamy grey podzolics on the lower levels. Under natural vegetation there is rapid run-off on the steeper slopes. Table 6.1 gives some indication of the soil erosion problem for whole northern region of Thailand.

Table 6.1.
Soil Erosion in Northern Thailand

Extent of Erosion	Area Eroded ('000 rai)	Percentage of Total Area
Moderate	5,151	4.85
Severe	17,983	16.92
Very Severe	6,141	5.78
Total	29,275	27.55

Source: Attaviroj (1986)

Estimates for the total HASD project area are not available, but project staff have produced experimental data for sites used in traditional farming. Hoey et al. (1987) have estimated that under traditional farming practices, the annual soil loss is 45 t/ha. If this estimate were consistent across all traditionally cultivated land in the HASD area (100,000 to 230,000 ha) the total soil loss would range from 4.5 to 10.35 million t/y.

Traditional Farming

The predominant system in the highlands has been a sedentary paddy-based system involving a single crop of wet season rice with minimal or no second cropping and a swidden or shifting system in the hills involving a single crop of wet season upland rice or maize. Swidden agriculture is the traditional method of hill tribe farming, involving cutting and burning forest, growing a crop for one or two years and then moving onto another area and repeating the process. Each area was traditionally bush fallow for 10-15 years between cropping - a period sufficient to maintain soil fertility and crop yields. Hoey et al. (1987) have reported that by 1987 the average bush fallow period had been reduced to less than 4 years, although it varied greatly, especially in some areas where annual cropping had replaced swidden. Increasing population and reclaiming of land by the Thai government for forestry were the reasons for this significant reduction in the swidden cycle. A fallow period of 4 years is not considered sufficient time to replenish nutrients and organic matter. As a consequence, yields are declining and soil erosion is increasing. Hoey et al. (1987) suggest that the trend of decreasing fallow periods is likely to continue. They state:

> "the diminishing fallow period, declining crop yields, increasing rice deficiency, poverty, soil erosion and deforestation will accelerate and lead to ecological and social disaster areas in the north, unless changes are made."

Rice is the food crop and grown as paddy or as upland rice. Maize is the cash crop. Hoey et al. (1987) report that at the commencement of the HASD project, in some places excess rice was sold for cash, especially where dry season rice was being grown. Opium and garlic appeared to be the only cash crops of significance. The same authors found that yields in the paddy and the uplands were very poor, with yields of less than 0.5 t/ha for upland rice in some areas. In many villages not enough rice was produced to eat. In addition to the effects of decreasing soil fertility on yields, crop husbandry was less than optimal. In particular, infestation by weeds was a problem.

Whereas once the hill tribes could have coped with the problems of decreasing yields and increased population growth by moving to new virgin land, this was no longer possible by the early 1980s. There was no further farming land available in Northern Thailand. In fact, hill tribe farmers were losing land to forestry. The only way of acquiring new land was by illegal encroachment upon forest areas.

Hoey et al. (1987) note that the contribution of soil erosion to the overall problem was poorly perceived, and consequently, no attempt was made by farmers to control the greatest single threat to their livelihood.

HASD Project Initiatives

The HASD project had to address both the immediate needs of the farmers for increased yields (the short term problem) and the long term sustainability of farming in the highlands. From an integrated environmental management perspective there was also a need to formulate policies addressing the serious off-site aspects of siltation and sedimentation.

A multiple-stage approach was adopted. Short term solutions based on the existing food and cash crops were developed and tried in a number of villages, the aim being to give immediate results. The next phase was the integration of these initiatives into new cropping patterns and systems. The idea was that cropping systems would become part of an integrated farming system encompassing field crops, tree crops and animals. A fundamental requirement was that the technology and systems being promoted should be compatible with both the short term and long term objectives.

The first step in the process was the introduction of a number of simple, effective initiatives which focussed on the underlying causes of poor yields, fertility decline and weed competition. It consisted of the application of fertilisers, line planting and weed control.

The next step was aimed at increased cropping intensity through multiple cropping and the introduction of soil conservation measures. Of prime importance was the requirement that these conservation measures would be applicable given the social, attitudinal and institutional situation in the highlands. Constraints included fragmented land tenure, piecemeal adoption of the initiatives (both temporally and spatially), a low knowledge base and a severe lack of financial resources. For the project to be successful it ultimately had to be accepted throughout the highlands and it had to be self-funding.

In the initial phase, based on existing food and cash crops, little or no contribution was made to fertility enhancement and/or erosion control. The extensive use of legumes, initially as individual crops and then as component crops of integrated patterns, was aimed at increasing soil fertility and the generation of additional cash income. Cash outlays for the use of legumes are relatively modest.

The next step was the introduction of erosion control measures. Various techniques were tested in the field by project staff. They included terraces, strip cropping, strip cropping between leucaena strips and strip cropping between grass strips. Each of these techniques (the with-project situation) was compared with the traditional farming system of rice monoculture (the without-project situation). Table 6.2 presents the results, in terms of gross margins, soil loss and sediment load in runoff, for the four systems of erosion control. The alternatives are ranked on the basis of financial viability and environmental cost, with soil erosion as the main indicator of environmental damage.

Table 6.2.
Results from the Upland Evaluation Sites for 1985 and 1986

Treatment	GM ($US/ha)		Runoff	Soil Loss	Av. Conc. Sediment in Runoff	Ranking		
	Maize	Rice	cu m	t/ha	kg/cu m	Returns	Soil Loss	Average
Traditional	200	55	2770	45.0	16.05	4	5	4.5
Terraces	139	7	610	1.7	2.76	5	2	3.5
Strip Cropping	203	103	880	5.2	5.84	3	3	3.0
Leuceana	259	63	1260	21.0	16.25	2	4	3.0
Grass Strips	275	95	513	1.3	2.49	1	1	1.0

Source: Hoey (1988)

Economic Evaluation of the HASD Project

General Approach

The framework for an economic evaluation of the HASD project is an environmental benefit-cost analysis. The term "environmental" as used here refers to ecological, sociological and economic impacts. The approach incorporates the same kinds of analyses that would be undertaken in the preparation of EIAs of development projects, but includes economic valuations of the relevant effects. In this case, the project is an integrated land use management and social development programme. Inclusion of off-site as well as on-site impacts is an important feature of the analysis.

The evaluation compares two situations: the without-project situation and the with-project situation. The analysis involves only hill-tribe households actually participating, or predicted to participate, in conservation farming during the period 1982 to 1995.

Without-Project Situation

To carry out a proper evaluation of the project, future conditions must be predicted. An accurate prediction even of the without-project situation is difficult. The base year for evaluation was assumed to be 1982. Realistically, this situation would change over time, notwithstanding implementation of the project objectives. Economic, social and institutional forces will inevitably alter the life-styles and land use practices of people cultivating the highlands. Increasingly, there are incentives for hill-tribe people to become involved in the cash economy, by producing vegetables, fruits and tourist artifacts for sale. Cash crops have become necessary to supplement the hitherto subsistence life-style and economy of many villages. Closer social contact with lowland Thais and the development of tourism are leading to an integration of the hill tribes into the Thai social fabric and economy. Other factors could also produce change, such as the Thai government's resolve to protect the remaining forests and reclaim areas for reafforestation.

The HASD project is a component of the overall program of the Hill Tribe Welfare Division of the Public Welfare Department, which has many other developmental features.

Objectives of the Project

The project objectives are: (i) to raise the standard of living of the rural poor in the highlands; (ii) to increase and sustain the production of subsistence crops (rice) and cash crops (maize, legumes, coffee, tea, fruit trees and bamboo); (iii) to reduce soil degradation (ie depletion of organic matter and decreased availability of phosphorous, nitrogen, potassium and important trace elements), eventually obviating the need for swidden agriculture; (iv) to provide health and education facilities for those not presently covered by such; (v) to assist in the integration of hill tribe villagers into Thai society; and (vi) to contribute to the training of staff in farm management, health, education and general areas of administration.

In terms of agricultural development the objectives are: (i) to reduce the area of steep land under cultivation; (ii) encourage reafforestation of degraded areas; (iii) improve land use patterns; (iv) introduce sustainable agricultural production techniques; and (v) encourage the replacement of opium with other crops in opium growing areas.

The With-Project Situation

The with-project situation involves the gradual adoption of conservation farming and social welfare improvement. From the commencement of the project in 1982, 5000 households participated. At present, 7700 households are involved, each with an average of six members. This is less than 10% of the hill tribe population in the highlands. In the future, 7700 households are still predicted to remain in the project. For these participating farmers, the project benefits and costs can be measured reasonably accurately.

The demonstration effect of the project could be very important. If those presently not involved can see the benefits of the project, they are likely to adopt conservation farming. With additional expenditure on the provision of extension services, the whole highland area and total hill tribe population could adopt conservation farming. The consequent benefits could be substantial.

Valuation of On-Site Effects

Assessing With-Project Effects

The on-site effects of the project are measured as the increase in the net value of production for the average hill tribe farm (or household). The with-project benefits must be estimated relative to the without-project situation. Relevant estimates are shown in Table 6.3 for years 1982 through 1995. Soil degradation causes the without-project baseline to shift through time. Before any estimates of net benefits could be made, changes in the baseline had to be established.

The HASD project team (Australian Agricultural Consulting and Management Company) has developed computer farm models which estimate the financial situation for the average participating household, with and without the project. Data from these models were used in the analysis. The models present the situation at three points in time: 1979 to 1982 (pre-project, represented as a steady state); 1987; and 1995. A steady state is assumed for participating farmers in 1995. Both the with- and without-project changes are assumed to occur linearly between the reference years.

Certain important features of the models need to be explained. Output prices are based on conservative estimates. All prices are shown in constant 1987 prices and prices of both inputs and outputs are held constant. Only minimal capital investment is involved in traditional farming and its opportunity cost is valued at zero. Additional capital is not required for the adoption of conservation farming.

Net income is calculated excluding labour costs. This assumes that the opportunity cost of hill-tribe farm labour is zero, a not unreasonable assumption, given the lifestyle and location of the hill tribe farmers and the economic situation prevailing in the highlands. There is some empirical evidence to support this assumption. For example, Jones (undated) reports that for a particular hill-tribe village located 46 km by road from Chiang Rai and containing almost 380 people, there were no opportunities for off-farm employment. Nevertheless, the with-project situation requires a 15% to 18% increase in the use of labour. Using an estimate of 30 Baht per day for labour costs derived by Attaviroj (1986) estimated net farm benefits would decrease by 11% to 14%.

Finally, in estimating the with-project situation, various adoption rates for conservation farming are built into the models. The average participating farm is the reference point.

Estimates of net benefits, before the application of discounting, are shown in columns 5 and 6 of Table 6.3. If the 1979/82 steady state continued for the 14 years 1982 to 1995, the total net household income for participating households would be 721,489,300 Baht, without discounting. The net loss of income resulting from soil degradation over the period 1982 to 1995 (inclusive) for the participating households can be calculated as 71,691,900 Baht. From 1982 to the assumed steady state in 1995, the project reverses the downward baseline trend in without-project income and raises by a substantial amount the net household income of participating farmers. The gain in net income over the 1979/82 steady state is 456,706,500 Baht. To estimate the total change from the without-project situation to the with-project situation, these two figures must be added. The net benefit, in undiscounted terms, is thus 528,398,400 Baht (see column 5 of Table 6.3) or Aust $26,419,920 (see column 6 of Table 6.3).

Calculation of Net Present Value

Net present value of the benefits of the project can be calculated in Aust $ by applying an appropriate discount rate to the values in column 6 of Table 6.3. Discount factors for a discount rate of 10% are shown in column 7, and present values in column 8. The net present value of the gain to all participating households is estimated at approximately Aust $9.4 million with a discount rate of 10%. For a discount rate of 5% the gain is Aust $15.4 million, and at 15% discount it is approximately Aust $6 million.

Table 6.3.
On-Site Net Benefits to Households, 1987 Values

Year	Without-Project Net H'hold Income	With-Project Net H'Hold Income	Increase in Net H'Hold Income	Increase for Participating H'Holds		Discount Factor 10%	Net Present Value
	Baht	Baht	Baht	Baht m	A$ m		A$ m
(1)	(2)	(3)	(4)	(5)	(6)	(7)	(8)
1980							
1981	0	0	0	0	0	0.9091	0
1982	7,651	7,651	0	0	0	0.8265	0
1983	7,595	8,339	744	3.720	0.186	0.7513	0.139
1984	7,539	9,027	1,488	7.440	0.372	0.6830	0.254
1985	7,483	9,715	2,232	11.160	0.558	0.6209	0.346
1986	7,427	10,403	2,976	14.880	0.744	0.5645	0.436
1987	7,371	11,091	3,720	28.644	1.432	0.5132	0.735
1988	7,193	11,754	4,561	35.119	1.755	0.4665	0.819
1989	7,014	12,418	5,404	41.610	2.080	0.4241	0.882
1990	6,835	13,081	6,246	48.094	2.404	0.3855	0.927
1991	6,657	13,745	7,088	54.577	2.728	0.3505	0.956
1992	6,479	14,408	7,929	61.053	3.052	0.3186	0.972
1993	6,300	15,072	8,772	67.544	3.377	0.2897	0.978
1994	6,121	15,735	9,614	74.027	3.701	0.2633	0.974
1995	5,942	16,400	10,458	80.963	4.026	0.2394	0.963
Total			71,232	528.398	26.419		9.386

At this point a comparison can be made between the net present value of the project's on-site benefits and on-site costs, to produce an on-site benefit-cost ratio. The project's costs are presented in Table 6.4. Both Australian aid funds and contributions by the Thai government are included, with the first outlays occurring in 1980. Present value calculations for costs are shown using a discount rate of 10%. The result is approximately Aust $8.7 million. The net present value of costs is Aust $10.8 million at 5% discount, and Aust $7.2 million at 15%.

The on-site benefit-cost ratio for participating households is clearly sensitive to the discount rate. At 5% discount the ratio is 1.43; at 10% it is 1.08; and at 15% it is 0.83. This sensitivity is the result of higher costs occurring in earlier years and higher benefits in later years.

Table 6.4.
Project Costs, Australian Aid and Thai Government Contribution

Year	Baht m	Aust $ m	Discount Factor 10%	Net Present Value Aust $ m
(1)	(2)	(3)	(4)	(5)
1980	10.954	0.547		0.547
1981	11.207	0.560	0.9091	0.509
1982	24.837	1.241	0.8265	1.026
1983	32.367	1.618	0.7513	1.215
1984	32.918	1.645	0.6830	1.124
1985	38.120	1.906	0.6209	1.183
1986	40.430	2.021	0.5645	1.141
1987	47.436	2.371	0.5132	1.217
1988	23.891	1.194	0.4665	0.557
1989	2.550	0.127	0.4241	0.054
1990	1.470	0.073	0.3855	0.028
1991	1.470	0.073	0.3505	0.025
1992	1.470	0.073	0.3186	0.023
1993	1.470	0.073	0.2897	0.021
1994	1.470	0.073	0.2633	0.019
1995	1.470	0.073	0.2394	0.017
Total	273.530	13.676		8.712

Evaluating Demonstration Effects

A demonstration effect is likely to occur as a consequence of the project, with increasing numbers of hill-tribe households adopting conservation farming beyond 1995. While additional costs, primarily in providing extension services, would probably be required, the net benefits in these future years would be relatively high. It should be noted that less than 10% of the potential population of hill tribe farmers will be involved by 1995. Uncertainties and insufficient data prohibit extending the analysis beyond 1995 in other than an illustrative way. This analysis is carried out in a later section.

Valuation of Off-Site Benefits

General Nature of Off-Site Benefits

Conservation farming leads to various types of off-site benefits. They consist largely of the value of the damage prevented to other productive resources. This involves using the productivity changes approach to economic valuation of off-site benefits. To value off-site effects, the with- and without-project situations must be determined. The difference between the two is the net off-site benefit.

Most off-site benefits result from the reduction of soil erosion and its consequent downstream effects. Soil erosion leads to siltation and sedimentation of streams and water storages. The deposition of sediment in dams and reservoirs depletes their storage capacity and reduces the availability of water for irrigated crops. Deposition can eventually reduce the efficiency of hydro-electricity schemes through a loss of generating capacity and increased physical depreciation of generating equipment. It can raise the frequency and peak levels of flooding, increase dredging costs and reduce the efficiency of water traffic. With increased turbidity and higher nutrient loads, algal blooms may occur, adversely affecting freshwater fisheries.

The other major off-site benefit of conservation farming is forest protection. As conservation farming drastically reduces the dependence on swidden agriculture and the incentive to clear forests, there will be a reduction of erosion in forest areas and a potential increase in fuelwood and timber yields.

The analysis of off-site effects parallels the assessment of on-site benefits by confining the discussion to hill-tribe farmers participating in the project. As will be seen, this results in relatively insignificant off-site benefits. It is only if conservation farming is adopted throughout the highlands that very significant off-site benefits are likely to occur.

Sedimentation and Siltation

One of the major off-site effects of adopting conservation farming is a reduction in sediment deposited in downstream dams and reservoirs, as a consequence of reduced soil erosion. Sediment not transported directly to water storages, or a considerable portion of it, is deposited in streams and on low-lying land.

Over time, stream bank erosion and tributary erosion contribute to the sediment in the major water storages. Hence, the extent of annual water storage sedimentation is a function of annual soil erosion in the upstream watershed, stream bank or channel erosion, tributary bedload and the sediment delivery ratio. The volume of sediment accreting in a water storage does not necessarily reduce the useful storage capacity by the amount deposited, as a portion of the total capacity is "dead" storage which cannot be used for irrigation, hydroelectricity generation or other purposes.

Insufficient data exist to estimate the sediment transport process and the useful storage capacity of reservoirs affected by highland soil erosion. It can be assumed, however, that storage capacity is reduced to the full extent of sediment deposited. This results in an over-estimation of the effects, but the results are considered to be indicative of the impacts that are caused. The deposition contributed by participant farmers is small compared with total deposition from all sources, but with demonstration effects on other farmers a more significant difference could be expected.

Effects on Agriculture

Water storage depletion reduces the availability of water for irrigated crops. The estimates here are based on a similar study by Attaviroj (1986). Water allocation of 2500 cu m/rai (15,625 cu m/ha) is the necessary annual allocation for the second crop of rice. With reduced water supplies, the net income lost is 552 Baht/rai (3450 Baht/ha). It is assumed that the irrigable land in question has no economic use without irrigation. The estimate is based on average yields and prices over the period 1979/80 to 1983/84 and is probably an underestimate in terms of 1987 prices.

Certain assumptions can be made to evaluate the effect of soil loss from the activities of the participating hill tribe farmers. The changes between the reference points (1979/82, 1987 and 1995) are linear; there is a discrete increase in the number of participating farmers in 1987; and the area of land (and therefore the volume of soil erosion) under the differing management regimes - traditional farming and conservation farming - increases throughout the period but at marginally different rates. Finally, sediment expressed in terms of weight (tonnes) is converted into volumes (cubic meters) by assuming that one tonne of sediment equals 0.7 cu m.

Conservation farming reduces soil loss from 45 t/ha/y to 1.3 t/ha/y. Table 6.5 shows the estimated annual reduction in damage costs under the stated assumptions. They are small relative to the on-site benefits of the project. As seen in the table, the present value of net benefits is Aust $14,856 with a discount rate of 10%. At a rate of 5% the NPV is Aust $24,152 and at 15% it is Aust $9,548.

Table 6.5.
Damage Costs Avoided: Offsite Agricultural Production

Year	W/out-Project Baht	With-Project Baht	Damage Costs Avoided Baht	Damage Costs Avoided Aust $	Discount Factor 10%	Net Present Value Aust $
(1)	(2)	(3)	(4)	(5)	(6)	(7)
1980						
1981					0.9091	0
1982	83,462	83,462	0	0	0.8264	0
1983	92,471	85,582	6,889	344	0.7513	258
1984	101,480	86,377	15,103	755	0.6830	516
1985	110,488	85,759	24,729	1,236	0.6209	767
1986	119,497	83,816	35,681	1,784	0.5645	1,007
1987	120,557	80,548	40,009	2,000	0.5132	1,026
1988	121,528	70,921	50,607	2,530	0.4665	1,180
1989	122,500	61,382	61,118	3,056	0.4241	1,296
1990	123,560	51,756	71,804	3,590	0.3885	1,384
1991	124,531	42,129	82,402	4,120	0.3505	1,444
1992	125,503	32,590	92,913	4,646	0.3186	1,480
1993	126,563	22,963	103,600	5,180	0.2897	1,501
1994	127,534	13,336	114,198	5,710	0.2633	1,503
1995	128,506	3,709	124,797	6,240	0.2394	1,494
Total						14,856

Effects on Hydro-Electricity Generation

Deposition in the hydro-electricity producing dams in Northern Thailand is predicted to reduce the capacity of these dams to provide hydro-power. While the lack of empirical data precludes accurate assessment of forgone electricity generation capacity, it can be assumed that capacity is lost in proportion to the rate of sedimentation. Any additional maintenance costs of generating equipment are omitted.

Attaviroj (1986) has estimated the loss of hydro-power capacity in Northern Thailand dams from sedimentation to be 1.706 million kW per annum, which translates into 2.559 million Baht annually if valued at the average cost to consumers. It is not clear whether Attaviroj took into account the effective storage capacity (the capacity above the "dead" storage pool).

The impact of hill-tribe farmers participating in conservation farming is minimal. On the basis of 45 t/ha/y of soil erosion from traditional farming practices, applied to the area farmed by the participating farmers during the period 1982 to 1995 inclusive, the sediment deposition would be only 2% to 3% of total deposition. The without-project situation is evaluated on this basis. Estimates of off-site benefits (damage costs avoided) in electricity generation as a result of the project are presented in Table 6.6. At a 10% discount rate the present value of net benefits is Aust $9,474. With a rate of 5% the NPV is Aust $15,407 and a rate of 15% gives an NPV of Aust $6,088.

Fisheries Production

The water quality in both the major storages (the Bhumibol and Sirikit dams) was considered in 1987 to be suitable for aquatic life. It would seem, therefore, that the potential impacts leading to a reduction in fish harvests or an increase in the cost per unit effort had not occurred by then. Insufficient data are available to predict any adverse effects on fisheries production that might result from increases in turbidity and nutrient loadings in streams leading to the dams.

The gross value of fisheries production in the Bhumibol and Sirikit dams is substantial: approximately 19 million Baht per annum, excluding aquaculture. Any damage avoided by the introduction of conservation farming would have to be included as a benefit in the off-site impact analysis. This benefit must remain unquantified, given the lack of relevant information.

Table 6.6.
Value of Foregone Hydro-Power and Damage Costs Avoided

Year	W/out-Project	With-Project	Damage Costs Avoided		Discount Factor	Net Present Value
	Baht	Baht	Baht	Aust $	10%	Aust $
(1)	(2)	(3)	(4)	(5)	(6)	(7)
1980						
1981					0.9091	0
1982	53,313	53,313	0	0	0.8264	0
1983	59,512	54,447	5,065	253	0.7513	190
1984	63,975	55,630	8,345	417	0.6830	285
1985	71,083	54,447	16,636	832	0.6209	517
1986	75,265	53,313	21,952	1,098	0.5645	620
1987	77,545	51,180	26,365	1,318	0.5132	676
1988	77,545	44,895	32,650	1,632	0.4665	761
1989	77,545	39,369	38,176	1,909	0.4241	810
1990	77,545	32,808	44,737	2,237	0.3855	862
1991	79,969	26,937	53,032	2,651	0.3505	929
1992	79,969	20,637	59,332	2,966	0.3186	945
1993	79,969	14,623	65,346	3,267	0.2897	946
1994	82,548	8,587	73,961	3,698	0.2633	974
1995	82,548	2,394	80,154	4,007	0.2394	959
Total						9,474

Degradation of the Drainage System and Flooding

Sedimentation in the major storages in Northern Thailand reduces the capacity of storages to prevent flooding. Flooding can cause damage to agricultural crops and to urban areas. Remedial work, in particular dredging, can alleviate the problem, but at a cost. Dredging downstream areas may be necessary to permit boat traffic and shipping.

Bhumibol and Sirikit dams have played, and continue to play, a noticeable role in downstream flooding but the actual benefits are still in question. Data are not available by which to distinguish the effects of the dams from the other factors that contribute to flooding. Presumably any reduction in sedimentation resulting from the adoption of conservation farming in the highlands would provide marginal benefits. At this point in time they have not been quantified.

Attaviroj (1986) calculated the cost of dredging 19 million cu m of sediment from the first 18 km of the Chao Phraya River as 424 million Baht in one year. The sediment was attributed to exploitive monoculture in the Northern Thailand uplands. This was the most significant off-site cost identified by Attaviroj. It had a marked effect on the results of his evaluation.

While it is self-evident that soil erosion resulting from traditional farming practices by participating farmers contributes to the need for downstream dredging, it is not possible to estimate its full extent. Confined to participating farmers in the HASD project the effect would be marginal. The total effect of traditional farming throughout the highlands, however, could be substantial. Much more must be learned about the sediment transport process throughout Northern Thailand and the Chao Phraya River, including the deposition and retention of sediment on low-lying land and in water storages, before meaningful attempts can be made to estimate the downstream benefits of conservation farming.

Forest Protection

Forest protection is a positive benefit of conservation farming. If conservation farming practices were adopted, the area farmed by each household could be reduced by two to four times and still keep the household supplied with staples (AACM 1987). This would significantly reduce the incentive to encroach upon forest land and allow surplus land to be returned to forest.

With the continuation of traditional farming, forest areas would increasingly be encroached upon by hill tribes. If the need for extra farming land without the project were assumed to double over the period 1982 to 1995, it is a straightforward task to estimate the gain in forest area that could be attributed to the project. This gain can then be converted to a difference in timber yield, assuming that the timber is actually harvested. (If the Thai government decided not to allow commercial forest production in this higher elevation area, for environmental or any other reasons, a more complex analysis would be required to measure the value of protected forest.)

It is assumed that, under conservation farming, 15 rai with a cropping intensity of 131% can provide self-sufficiency, while swidden agriculture requires 30 rai. The gain in forest land and the net benefits of extra timber yields can be calculated by taking the difference and multiplying by the number of participating households. The value of timber production in Northern Thailand is assumed to be 1350 Baht per rai. The results of these computations are shown in Table 6.7. The NPV of forest protection is Aust $18,695,095 with a discount rate of 10%, Aust $30,561,175 at 5% and Aust $11,929,626 at 15%.

Table 6.7.
Economic Value of Gain in Forest Area and Damage Costs Avoided

Year	Area	Annual Gain		Discount Factor 10%	Net Present Value
	rai	Baht m	Aust $ m		Aust $ m
(1)	(2)	(3)	(4)	(5)	(6)
1980					
1981				0.9091	0
1982		0	0	0.8264	0
1983	5,750	7.762	0.388	0.7513	0.291
1984	11,500	15.525	0.776	0.6830	0.530
1985	17,250	23.287	1.164	0.6209	0.722
1986	23,000	31.050	1.552	0.5645	0.876
1987	44,275	59.771	2.988	0.5132	1.533
1988	53,130	71.725	3.586	0.4665	1.672
1989	61,985	83.679	4.183	0.4241	1.774
1990	70,840	95.634	4.781	0.3855	1.843
1991	79,695	107.588	5.379	0.3505	1.885
1992	88,550	119.542	5.977	0.3186	1.904
1993	97,405	131.496	6.574	0.2897	1.904
1994	106,260	143.451	7.172	0.2633	1.888
1995	115,500	155.925	7.796	0.2394	1.886
Total					18.695

Comparison of Benefits and Costs

The economic benefits and costs of the HASD project can now be compared, taking into account both on-site and off-site effects, remembering that the evaluation is confined to participating farmers, only 10% of the total population of hill tribe farmers. The comparisons in terms of net present values using discount rates of 5%, 10% and 15% are presented in Table 6.8.

Table 6.8.
Net Present Values: Participating Farmers, 1980 to 1995

		Benefits Aust $	Costs Aust $	Benefit-Cost Ratio
5% Discount Rate				
On-site		15,375,238		
Off-site				
	Irrigation	24,152		
	Hydro-power	15,407		
	Fisheries	not estimated		
	Drainage/Flooding	not estimated		
	Forests	30,561,175		
Total		45,975,972	10,778,316	4.2:1
10% Discount Rate				
On-site		9,386,659		
Off-site				
	Irrigation	14,856		
	Hydro-power	9,474		
	Fisheries	not estimated		
	Drainage/Flooding	not estimated		
	Forests	18,695,095		
Total		28,106,084	8,712,385	3.2:1
15% Discount Rate				
On-site		5,956,031		
Off-site				
	Irrigation	9,548		
	Hydro-power	6,088		
	Fisheries	not estimated		
	Drainage/Flooding	not estimated		
	Forests	11,929,626		
Total		17,901,293	7,198,061	2.5:1

The inclusion of off-site effects, in particular the very substantial gain in commercial forest production, makes a significant difference to net benefits and the benefit-cost ratios. The ratios are positive and large for all three discount rates. It may be recalled that at a 10% discount rate the benefit-cost ratio for on-site effects was 1.08. It increases to 3.2 when the off-site benefits are included. The analysis reveals that, even without taking into account the potential demonstration effect, with more hill tribe farmers adopting conservation farming, the HASD project results in a significant improvement in community economic welfare.

Adoption of Conservation Farming

The average household income with-the-project is predicted to be 16,400 Baht in 1995. The financial benefits to participating farmers are very likely to have a major demonstration effect throughout the hill tribe population of Northern Thailand. With only 7,700 farming households participating in 1995, from a total of 83,333 households, an additional 75,633 farming households could adopt conservation farming. Even without the provision of information through extension work it is probable that non-participating households would see the advantages of conservation farming.

While it is not possible to predict the adoption rate, with or without the extension services, it is possible to illustrate the potential benefits by making arbitrary assumptions. For example, if all households adopted conservation farming by year 2010 (5000 households per year for 16 years), the annual incremental gain would be 53 million Baht or Aust $2.6 million. By the year 2010, the on-site gain would be Aust $39 million annually. The results of these computations, including net present values, are shown in Table 6.9. The results are clearly sensitive to the discount rate.

The nature of the adoption path will also affect the NPV. For the calculations in Table 6.9 a linear adoption path is assumed. A lower adoption rate would result in smaller economic benefits, while a faster rate would produce larger gains.

Off-site Costs at the Regional Scale

As already stated, the contribution of farmers participating in the HASD project to soil erosion and its off-site costs is small compared with regional off-site effects. Potentially, the increase in off-site costs could be ten-fold or more. Such costs, measured in terms of lost production, are occurring now and they will continue unless there is rapid adoption of conservation farming throughout the highland areas.

Table 6.9.
On-Site Benefits With All Hill Tribe Farmers Adopting Conservation Farming

Year	Annual Gain Aust$ m	Discount Rate 5%	NPV 5% Aust $ m	Discount Rate 10%	NPV 10% Aust $ m	Discount Rate 15%	NPV 15% Aust $ m
(1)	(2)	(3)	(4)	(5)	(6)	(7)	(8)
1995							
1996	2.6	0.4581	1.19	0.2176	0.57	0.1069	0.28
1997	5.2	0.4363	2.31	0.1978	1.03	0.0930	0.48
1998	7.8	0.4151	3.24	0.1799	1.40	0.0808	0.63
1999	10.4	0.3957	4.12	0.1635	1.70	0.0703	0.73
2000	13.0	0.3769	4.90	0.1486	1.93	0.0611	0.79
2001	15.6	0.3589	5.60	0.1351	2.11	0.0531	0.83
2002	18.2	0.3419	6.22	0.1228	2.23	0.0462	0.84
2003	20.8	0.3256	6.77	0.1117	2.32	0.0402	0.84
2004	23.4	0.3101	7.26	0.1015	2.38	0.0349	0.82
2005	26.0	0.2953	7.68	0.0923	2.40	0.0304	0.79
2006	28.6	0.2812	8.04	0.0839	2.40	0.0264	0.76
2007	31.2	0.2678	8.36	0.0763	2.38	0.0230	0.72
2008	33.8	0.2551	8.62	0.0693	2.34	0.0200	0.68
2009	36.4	0.2429	8.84	0.0630	2.29	0.0174	0.63
2010	39.0	0.2314	9.02	0.0573	2.23	0.0151	0.59
Total	312.0		92.17		29.71		10.41

Policy Implications

The study suggests that the social gains far outweigh the costs of conservation farming, notwithstanding the fact that future gains must be discounted in an economic evaluation. The analysis presented in this case study strongly supports conservation farming practices, as opposed to traditional cultivation in the highlands. It has shown that the financial gains to farmers are large. These gains should provide the incentive for wide-scale diffusion of improved farming practices, but it is likely to take some time.

The study has shown that the off-site benefits, even for the small percentage of farmers participating, are significant. These benefits occur because other productive resources (forests, downstream irrigated farmlands, waterways and reservoirs) are protected from production losses. On a regional scale these will outweigh the financial benefits to farmers. While some losses could be prevented or minimised through the use of government regulations, as in the case of forest production, others such as sedimentation comprise negative environmental externalities that can be countered only at the source of the problem - the traditional farmer.

The case study suggests practical solutions for policymakers; and it also illustrates a method for the assessment and evaluation of environmental impacts. It does this by evaluating environmental impacts within the framework of benefit-cost analysis. While there are unresolved problems with the approach, such as the issue of intra- and inter-generational equity, it does provide a framework based on economic realities.

The second methodological lesson is that the study demonstrates the need to assess the with- and without-project situations, and especially a changing baseline over time. Only the net impacts should be considered. This procedure is often neglected in the assessment of projects.

Policymakers nevertheless should be aware of data deficiencies and assumptions underlying the analysis. In this study, the assumptions have been clearly stated. A project as potentially socially significant as the HASD project warrants detailed analysis; and that becomes possible only if sufficient, appropriate data are gathered on a continuing basis. It is a relatively inexpensive component of project planning and management - but so essential if the right decisions are to be made.

References

Arbhabhirama, A., D. Phantumvanit, J. Elkington and P. Ingkasuwan (1987) *Thailand Natural Resources Profile*, Oxford University Press, Oxford.

Attaviroj, P. (1986) 'Soil Erosion and Degradation: Northern Thai Uplands', paper presented to the International Conference on the Economics of Dryland Degradation and Rehabilitation, Canberra.

Australian Agricultural Consulting and Management Company (1987) 'Highland Agricultural and Social Development (HASD) Project: Summary of Financial and Economic Analyses', report to the Australian International Development Assistance Bureau, the World Bank and the Royal Thai Government Public Welfare Department.

Board of Investment (1992) *Investment Opportunity Study: Environment Markets in Thailand*, Bangkok.

Chaturabhad, P. (1988) *People of the Hills*, Third Edition, Editions Duang Kamol, Bangkok.

Hoey, M. (1988) 'Landuse, Related Problems and Development Strategies for Stable Agricultural Systems for the Highland Agricultural and Social Development Project', paper presented at ISCO, Bangkok.

Hoey, M. (undated) 'Conservation Farming Systems for the Highlands', Australian Agricultural Consulting and Management Company, Adelaide.

Hoey, M., S. Tuamcharoen and S. Tepsarn (undated) 'A Farming System Approach to Stable Agriculture in the Uplands of Northern Thailand', paper presented at the Steeplands Conference.

Hoey, M., S. Tepsarn and S. Thaumcharoen (1987) 'Toward Stable Agricultural Systems for the North Thailand Highlands', paper presented at the Farming Systems Seminar, Chiang Mai.

Jones, P. (undated) 'Village Profile - Mae Salap Village Mae Chan Zone', paper prepared for the Australian International Assistance Bureau and the Royal Thai Government Department of Public Welfare.

Kritiporn, P., T. Panayotou and K. Charnprateep (1990) *The Greening of Thai Industry: Producing More and Polluting Less*, from the Conference on Industrialising Thailand and Its Impact on the Environment, December 8-9, Chon Buri, Thailand. Research Report No. 5., Thailand Development Research Institute Foundation.

Panat, Dr (1983) 'Environment Legislation in Thailand', unpublished report, Thammasat University, Bangkok.

Panayotou, T. and C. Parasuk (1990) *Land and Forest: Projecting Demand and Managing Encroachment*, from the Conference on Industrialising Thailand and Its Impact on the Environment, December 8-9, Chon Buri, Thailand. Research Report No. 1., Thailand Development Research Institute Foundation.

Social Research Institute, Chulalongorn University (1987) 'Post Impoundment Evaluation and Development Planning of the Bhumibol and Sirikit Projects', report to the Electricity Generating Authority of Thailand, Vol 1, Bangkok.

Srikhajon et al. (1980) 'Soil Erosion in Thailand', Department of Land Development, Bangkok.

Walker, A.R. (ed.)(1986) *Farmers in the Hills*, Suvarnabhumi Books, Singapore.

Wanichsombat, C. (1988) 'Administration of Environmental Impact Assessment in Thailand', paper presented at Expert Group Meeting on Environmental Impact Assessment of Development Projects, ESCAP, Bangkok.

Chapter 7

SOCIAL FORESTRY IN NEPAL: ECONOMIC APPRAISAL OF RESOURCE MANAGEMENT

NEIL BYRON AND DAVID GRIFFIN

In undertaking any environmental impact assessment, it is important to predict what is likely to happen *without* the project as well as *with* the project. It is commonly asserted that massive soil erosion in Nepal is caused by human activity in catchments, especially deforestation, with downstream effects extending to India and Bangladesh, where heavy sedimentation and flooding occur. The suggestion is that investments in catchments would yield large environmental benefits in Nepal and other countries.

The case study presented here is the Nepal-Australia Forestry Project (NAFP), a bilateral project of His Majesty's Government of Nepal (HMGN) and the Australian International Development Assistance Bureau (AIDAB). Griffin (1988) has given a full description of the Project, including its gradual evolution over two decades of its operations, policy and conceptual framework. Only a summary of these matters can be given here.

In this case study, experts from the Australian National University working on NAFP contradict the accepted view on land degradation in Nepal. While the Project has yielded significant on-site benefits, the off-site benefits have been considered negligible. The authors conclude that investments directed towards the achievement of off-site benefits are not likely to give high economic returns. This finding may not be applicable in other countries or under other geomorphological conditions.

D. James, The Application of Economic Techniques in Environmental Impact Assessment, 183–216.

Background to Case Study

General Setting

Nepal is an independent kingdom in the Himalayas occupying a total area of 140,640 sq km, being 840 km in length and 226 km in breadth. It is divided into five regions (see Figure 7.1.)

The population of Nepal is approximately 16 million. The country's political structure consists of a democratic monarchy introduced by a constitution in 1951. The initial multi-party parliamentary system of government was found to be unworkable and a partiless Panchayat system was established in 1961 under the direct leadership of the King. The new 1961 Constitution provided for a national assembly as the legislative body, an independent judiciary and a council of ministers as the executive branch of government. In recent times the political system has become more democratic, after a popular revolt against the partiless panchayat system.

The country's economic and social development has been modernised through a series of 5-year plans, the first one commencing in 1956. These plans have been directed towards ending poverty, raising living standards and meeting the basic needs of the people. However, the overall economic performance of the country has remained low. This is related to the high rate of population growth and the low growth rate of gross national product. Around 90% of the population is still engaged in the agricultural sector which generates some two-thirds of the country's GDP. About 70% is represented by food grains and the rest by cash crops. Other natural resources are hydro-electric power, minerals and forests. The industrial sector represents approximately 5% of GNP and employs about 1% of the labour force.

Land Degradation in Nepal

The Popular View

The literature on land degradation and management in the Nepalese Himalaya is very large. Here it suffices to highlight major issues and point the reader to key publications. Both the popular and much of the scholarly writing on Nepal over more than a decade depicts the country as being in the grip of a severe ecocrisis. Important expressions of the view are to be found in Eckholm (1975) and The

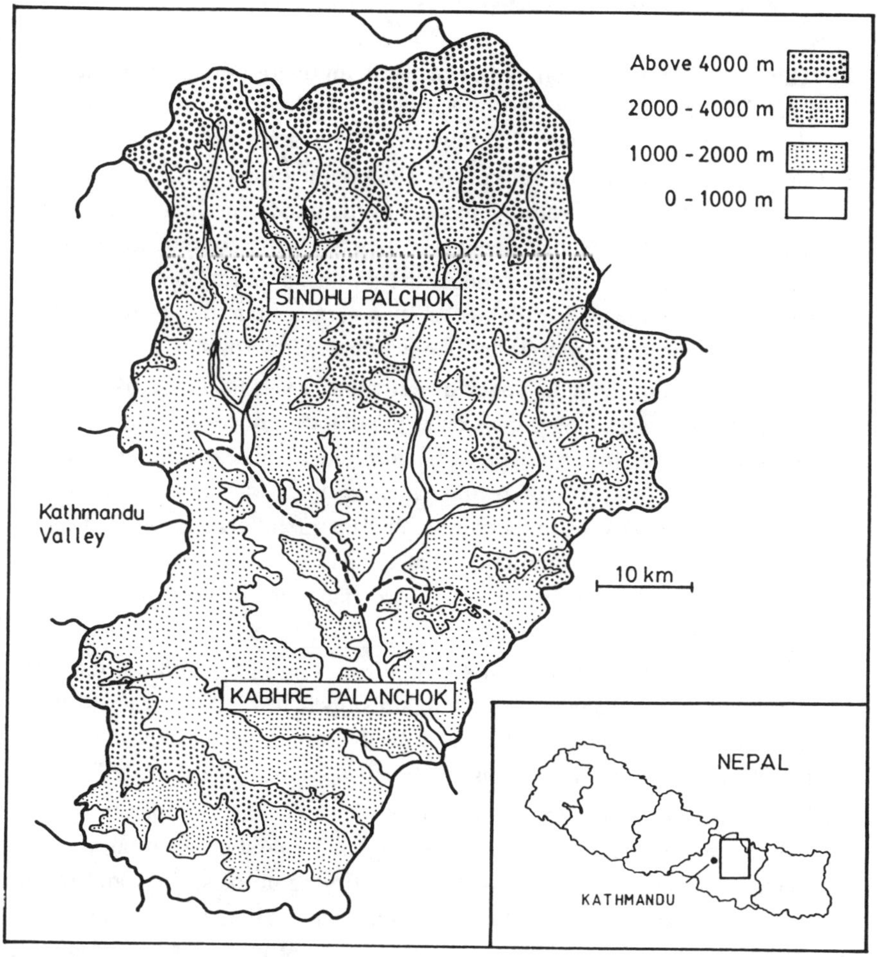

Figure 7.1. Map of Nepal Australian Forestry Project area.

World Bank (1978). Key components in the theory of Himalayan environmental degradation have been summarised by Ives and Messerli (1987) as follows:

1) "Following the introduction of modern health care, medicine, and malaria suppression in the Terai after 1950, there occurred an unprecendented wave of population growth that does not yet appear to have peaked. For Nepal as a whole it appears to have reached 2.5 per cent per annum for the 1971-1981 census decade (Goldstein et al. 1983) but in many cases it exceeds 3.0 - 3.5 per cent per annum. Nepal's total population today is probably in excess of 16 million. Comparative

population growth rates appear to be characteristic of the entire Himalayan region."

2) "This veritable population explosion, with an overall doubling period of about 27 years, is augmented by uncounted and uncontrolled illegal immigration from India into the Nepalese Terai across the open frontier. Furthermore, over 90 per cent of the 1981 population is rural and subsistence. This has led to rapidly increasing demands for fuelwood (more than 90 per cent of Nepal's energy depends on the combustion of biomass), construction timber, fodder (the domestic animal population has undergone an increase parallel to, or even greater than, that of the human population), and agricultural land on which to grow food."

3) "The next step in what has been described as a vicious circle, is that the needs of the burgeoning subsistence population are exerting increasing pressures on the forest cover. This has led to massive deforestation, amounting to a loss of half of the forest reserves of Nepal within a 30-year period (1950-1980) and comparable losses in large areas of the Indian Himalaya, and a prediction that by AD 2000 no accessible forest cover will remain; it is widely assumed that this situation has reached crisis proportions."

4) "The deforestation, accompanied by the cutting of agricultural terraces on steeper and more marginal mountain slopes, has led to a catastrophic increase in soil erosion and loss of land through accelerated landslide incidence, and to the disruption of the normal hydrological cycle."

5) "This, in turn has led to increased runoff during the summer monsoon and an increase in disastrous flooding and massive siltation on the plains, and to lower water levels and the drying up of springs and wells during the dry season. Related ills are: rapid siltation of reservoirs; abrupt changes in the courses of rivers; spread of barren sand and gravel across rich agricultural land on the plains; and increased incidence of disease in downstream areas."

6) "The increased sediment load of the rivers emanating from the Himalayan system is extending the Ganges and Brahmaputra delta and causing islands to form in the Bay of Bengal. Amongst the evidence cited are the extensive plumes of sediment that can be seen on Landsat imagery to extend several hundred kilometres into the bay."

7) "The continued loss of agricultural land in the mountains leads to another round of deforestation to enable the construction of more terraces on which to grow subsistence crops. Yet, as the labour of walking greater distances from the village to fuelwood supplies increases with the receding forest perimeter, a critical threshold is reached whereby the available human energy (principally female) becomes progressively over-taxed and an increasing quantity of animal dung is used for fuel."

8) "Consequently another vicious circle is linked to the first one: terraced soils are deprived of natural fertilizer - the animal dung now being used for fuel. This lowers crop yields. Also, the ensuing weakened soil structure further augments the incidence of landslides. Even more trees are cut on more marginal and steeper slopes to make room for more agricultural terraces to feed the ever growing subsistence population."

These views of the situation were highly influential during the 1970s: the original impetus for NAFP was founded in them. The challenge to them at first arose only gradually because of the apparently convincing nature of the ecocrisis theory. It accorded well with the perceptions of most short-term visitors to Nepal and with the views of the growing environmentalist pressure groups throughout the world. To this day it predominates in the popular media and survives in the scientific literature (Myers 1986).

The Revisionist View

Doubters of the popular view were at first tentative. Their doubts usually arose from inconsistencies between the facts as they knew them in an often restricted field or area and those which would be required to sustain the accepted theory of Himalayan environmental degradation. An opportunity to assess the evidence was created by the Mohonk Mountain Conference on the Himalaya-Ganges Problem held in 1986, and largely reported in Mountain Research and Development Vol.7, No.3 (1987) and Ives and Messerli (1989). This conference produced ample evidence of profound factual uncertainty, sufficient to raise doubt about the validity of the standard theory.

There are two major aspects of the "revisionist" view which are particularly relevant to our present purpose. The first relates to erosion. The second is based on history.

Empirical Studies of Soil Erosion

It now seems certain that geologic erosion in the Himalaya occurs on a far greater scale than that which can be attributed to the sum of human activities (Carson 1985; Gilmour 1986; Hamilton 1986, 1987; Byers 1987a,b; Bremmer 1987; Gilmour et al. 1987; Sattaur 1987). Remote downstream effects are almost entirely linked to geologic erosion. This is not to say that man-induced erosion does not cause important local effects. It does, mainly through the loss of top soil from agricultural and severely over-grazed areas and through the more obvious deep erosion of the clay soils widespread in the major valleys at moderate elevations in the Middle Hills. Extensive down-stream siltation and major flood events, however, are not man-induced and it is doubtful if they can be controlled by man.

Some details from Gilmour's (1985) review of the impacts of reforestation on hydrological values in Nepal, are instructive in this context. Gilmour observes there is a critical shortage of accurate measurement data on this subject in Nepal: by implication little can be substantiated at this stage. Gilmour refers to the work by Pandey et al. (1983) and notes that in the wide variety of sites those authors studied (including heavily degraded areas) total monsoon runoff varied only from 6.6 mm to 9.0 mm between sites - equivalent to only 0.55% of rainfall. This difference is not, in Gilmour's view, sufficient to produce the "catastrophic floods" downstream suggested by the authors.

Gilmour suggests that the protective role of forests in hill environments is important, but that the case should not be overstated. From the evidence available, he concludes that surface erosion and some minor landslips will be controlled to some extent by reforestation, whilst water yields from catchment and into springs may be reduced (although Gilmour suggests downstream effects of reduced catchment yield and flood flows will be minor). He also suggests that reduction of downstream sedimentation from reforestation will be insignificant, except on a very local scale, when compared to the impact of natural geological erosion. Finally, those who expect forestation or other human activity to affect the rate of sediment deposition in Bangladesh have apparently overlooked the time-lag between initial erosion in the hills of Nepal and final deposition in the Bay of Bengal, which could be centuries.

According to Sattaur (1987):

> "There is no statistically valid evidence to suggest that flooding and sediment production is any more or less than they were 200, 2,000 or even 2 million years ago. Rather, soil scientists believe that the sedimentation was, and still is, more a result of natural or geologic erosion."

Historical Analysis of Deforestation in Nepal

The second major revision of the ecocrisis view finds its strongest expression in the publications by Mahat et al. (1986a,b; 1987a,b) and Griffin et al. (1988). These authors studied human impact on the forests for Sindhu Palchok and Kabhre Palanchok, two districts in the Middle Hills of central Nepal, and in particular sought to determine causation in relation to rate of deforestation. The main conclusion, drawn from the districts but probably valid for the Middle Hills of most of central and eastern Nepal, is that serious deforestation is no modern phenomenon dating from the middle of this century. Griffin et al. (1988) have stated:

> "Rather, we affirm that the state of Nepal's (hill) forests arises... as a consequence of gradual evolution over centuries from petty states to the present kingdom."

Modern Nepal originated from the expansion of Gorkha under Prithivi Narayan Shah, commencing in 1743. A large army was created and also a large number of courtiers who expected recompense for their support. This led to a wave of deforestation, because forest was held in low regard: it did not have the same subsistence value as agricultural land, and, more importantly in the eyes of decision-makers, it did not yield taxes for the state as agriculture did. The peasant who farmed land was required to pay half the yield from it to the owner (or to the person who held the rights to such revenues, if not actual land title). Because of this heavy impost, many farmers preferred to clear new land because in doing so they were exempt from tax on their new land for a number of years.

Nearly all land with reasonable agriculture potential was already cleared by the turn of the century, and from then on the tax incentive system led to extensive degradation. This was exacerbated by the so-called rakam obligations - a form of tax in kind on landholders ostensibly for the state, but in reality often paid to powerful individuals. Payment was often in the form of charcoal, fuelwood, or iron, the latter manufactured in a very energy intensive manner in rural areas.

Political changes in 1951 weakened the authority of central government. The only significant legislation at this point affecting forestry was retrogressive: the Ailani-Parti law, which allowed treeless areas on the periphery of forests to be registered as private land. Again, the purpose was to broaden the taxable land base. One effect would have been to ensure that no re-planting occurred on these areas. Much of this treeless land proved unusable for agriculture, and became wasteland.

In essence, Mahat et al. (op cit) argue that the demarcation between forest land and agricultural land was established long ago. Their research suggests that within human memory in the two Districts, virtually no forest has been converted to grassland or agriculture - albeit that forest quality has declined under fuelwood and forage demand pressures. Much recent evidence, apart from Mahat et al. (1987a,b), indicates that the forest area in the Middle Hills has changed little in recent decades (Gilmour and Fisher 1991). Instead, forest quality has deteriorated as the rates of use exceed those of growth (Pearce et al. 1990). Hrabovszky and Miyan (1987) have projected that the cultivated area in neither the Middle Hills nor mountains will increase by more than 10% in the 20 years after 1985, with a consequent conversion of less than half that percentage of forest and grassland.

The same studies of Mahat et al. reveal that the causes of past deforestation are complex and that some are no longer operating although they were important in past times. To reverse the long-term trend will be no easy matter and will not occur through "quick-fix" methods. As explained by Griffin et al. (1988):

> "We also reject views that the present forest situation is redeemable solely by technological action or that it arose from peasant ignorance or mismanagement, from simple overpopulation, or through any loss of primitive purity in the face of modern market economics. The concept of 'the pressure of production on the resource' is found to provide a useful starting-point for the construction of a simple classification of the results of various pressures on the forest resource."

These historical reflections shed light on how the Middle Hills region of Nepal can be chronically short of forest, whilst displaying no evidence in recent years of widespread deforestation. More recent legislation which allocates forests to local government and community groups is in the process of replacing an incentive to deforest with an incentive to afforest, to rural dwellers.

Political and Institutional Setting

(Ben Boer)

In recent years the Kingdom of Nepal has developed a more democratic political culture, which has had a marked effect on the ability of government to address issues of pressing concern. The National Conservation Strategy of 1987, developed by the National Planning Commission in collaboration with the Nepal office of IUCN-the World Conservation Union, has had a good deal of influence in developing new policies on environmental impact assessment, environmental planning, the control of pollution and the protection of the national heritage.

The World Conservation Union has also sponsored a comprehensive study of the legal situation in relation to environmental and resource matters (NPC and IUCN 1992). That Report found that the regulation of environmental matters was in need of extensive overhaul. One of its recommendations was the drafting of comprehensive environment legislation to cover environmental planning, environmental impact assessment, pollution control and heritage conservation. An *Environment Protection Act* was drafted in 1992, which is expected to be considered by the Nepalese Parliament in 1993.

Environmental and Administrative Policy

(Ben Boer and Angus Martyn)

General Policy

In Nepal, there has been a growing awareness since the 1970s that natural resources should be conserved and managed on the principle of sustainable development (Shresta and Uprety 1988). The fifth Nepalese National Plan (1975-80) noted that development planning needed to be based on proper consideration of topographical, climatic and other environmental factors (NPC and IUCN 1992). Subsequent national plans during the 1980s recommended that increased attention be given to land-use planning, air and water pollution, forestry issues, soil conservation, the urban environment and environmental impact assessment. However, since these plans generally failed to provide specific and workable measures to implement these concerns, they had only very limited success in incorporating environmental priorities into national development planning (NPC and IUCN 1992).

In 1988, the Nepalese National Planning Commission, working in cooperation with the IUCN, published a National Conservation Strategy (NCS). The key objectives of the NCS were to:

1) Satisfy the basic material, spiritual and cultural needs of the people of Nepal, both present and future generations.
2) Ensure the sustainable use of Nepal's land and renewable resources.
3) Preserve the biological diversity of Nepal in order to maintain and improve the variety of yields and the quality of crops and livestock, and to maintain the variety of wild species, both plant and animal.
4) Maintain essential ecological and life-support systems, such as soil regeneration, nutrient recycling and the protection and cleansing of water and air.

As part of the NCS, a "Conservation Action Agenda" was developed with the aim of achieving these objectives in a manner compatible with the economic needs and the social and cultural values of the Nepalese people (NPC and IUCN 1992).

A new Constitution was proclaimed in the Kingdom of Nepal in 1990. The environment is addressed in Article 26(4):

> "The State shall give priority to the protection of the environment and also to the prevention of its further damage due to physical development activities by increasing the awareness of the general public about environmental cleanliness, and the State shall also make arrangements for the special protection of the rare wildlife, the forests and the vegetation."

The "duty" imposed upon the State by this Article is, however, contained within a section of the Constitution which is not susceptible to enforcement in a court of law. Nevertheless, it has the potential to force a greater degree of political accountability on the State concerning its policies on environment and development (NPC and IUCN 1992).

The lack of success during the 1980s in motivating the bureaucratic institutions towards effective environmental protection led to the establishment of the Council for the Conservation of Natural and Cultural Resources (CCNCR) in early 1991. (This Council is to be renamed and reconstituted as the Environment Protection Council under the proposed *Environment Protection Act*.) The Council is:

> "mandated to formulate policies on environmental planning and protection and is to arrange for both government and non-government participation in the implementation and monitoring of its policies. The Council has a multi-sectoral membership, with most of the key ministries, including finance, agriculture, forestry and soil conservation, industry, water resources, transport, housing and physical planning, education and culture, health, local development and tourism represented at secretary level. Provision is also made for CCNCR membership from the business community and non-government organisations." (NPC and IUCN 1992)

It has been proposed by the Council that structures should be created at the regional, district and Panchayat (local) levels to implement the NCS Conservation Action Agenda.

The Council was part of an attempt to adopt an approach that emphasises the building of constructive partnerships between the various government agencies, the private sector and local communities. The importance placed on public participation in developing, implementing and monitoring of environmental management programmes has been a feature of most recent Nepalese environment policy. Previously, little recognition had been given to the opportunities of unifying both the "centralised, bureaucratic - legalistic (government) system" with the traditional community-based system which had its own established networks and associations (NPC and IUCN 1992).

This new perspective has been continued in the drafting of the eighth Nepalese National Plan. For the first time a separate chapter on Environment and

Resource Conservation has been included. The major points outlined in this chapter include:

1) Identification of the overall impact on the environment of national policies, development projects, budget and investment related decisions;
2) The preparation, enactment and enforcement of comprehensive legal provisions which place priority on sustainability;
3) The explicit recognition of the crucial role of local units regarding local environmental planning and management systems;
4) An integrated and intersectoral approach in the development of environmental policies;
5) The development of pollution standards for individual areas and eco-systems;
6) The provision of financial incentives for the utilisation of the environment in a sustainable manner;
7) Preparation and execution of a National Environmental Action Plan;
8) The requirement for environmental impact assessment of proposed polices, programmes and projects.

Environmental Legislation and Administrative Guidelines

This section discusses the requirements for environmental impact assessment (EIA) under the National Environmental Impact Assessment Guidelines, as well as the provisions of the *Forest Act* of 1992.

In the decade to 1990, only 32 projects were subject to EIA (NPC and IUCN 1992). The majority of these assessments were initiated by foreign donor agencies. Although the seventh (1985-90) Nepalese National Plan made it mandatory for EIA to be introduced in relation to all major development projects concerned with tourism, water resources, transportation, urbanisation, agriculture and forestry, the lack of formal EIA procedures on either a legal or administrative level contributed to the failure in substantive implementation of this policy.

The adoption of the National EIA Guidelines by the Government in October 1992 is a major step towards the systematic incorporation of environmental considerations in development decisions. The guidelines are a product of cooperation between a wide range of Nepalese government agencies and the IUCN. They have been formulated in such a way to provide rigorous evaluation of development activities within the context of the institutional and human resource constraints of the Nepalese situation.

Only those activities that may have a significant effect on the environment are to be subject to formal EIA. The guidelines provide criteria on which a decision as to whether EIA is required should be based. These are:

1) The size, location, cost, output or infrastructure demands of the project;
2) The location of the project - will it affect environmentally sensitive areas?
3) The nature of the project - is it of a type that generally results in significant impacts?

The guidelines place emphasis on the role of public involvement in all stages of the EIA process, from evaluation of project feasibility to the monitoring and audit of the project. They note that the success of EIA largely depends on the participation and support of the affected community.

As the guidelines have been in place for only a short time, it remains to be seen how effective they are in promoting sustainable development. However, the apparent determination of the government to come to terms with environmental problems of Nepal indicate that EIA will be accorded a prominent place in future policy.

A second initiative relevant to this study has been the enactment of the *Forest Act* 1992. This is based on the Master Plan for the Forestry Sector 1988. The Act adopts a "resource-oriented", rather than the "use-oriented" approach typical of previous legislation (Belbase and Boer 1993). It divides forests into the categories of government-managed, protected, community, leasehold, religious and private forests and, with the associated *National Forest Rules* 1992, provides a basis for management of forests according to their particular designation. Again, because of the newness of the Act, it is not possible to evaluate its operation in the context of social forestry. Doubts have, however, been expressed in the recent past as to whether the decentralised, community-oriented approach to forestry policy is compatible with the attitude of Department of Forestry, with its "desire to retain ultimate control over forest policy... and to [intervene] where there is any diversion from its priorities" (Fisher 1989).

Description of Nepal-Australia Forestry Project

Land Use in the Case Study Area

Australian assistance to Nepalese forestry commenced in 1966. The Project is discussed here only as it has existed since 1978. From then it has operated in two districts in the Middle Hills of Nepal, in the Bagmati Zone of the Central Development Region. Griffin (1988) has described them as follows:

"Sindhu Palchok and Kabhre Palanchok Districts, presently with 175 panchayats, lie to the north-east, east and south-east of the Kathmandu Valley, extending from the main Himalayan range in the north to the valleys at the southern foot of the Mahabharat Lekh in the south. Elevation therefore falls from above 7,000 m to 300 m over a distance of only 100 km. ...It is an area of extremes, fertile paddy terraces contrasting with steeply sloping rain-fed terraces and areas of severe erosion. Although the total Project area is ca. 390,000 ha, only ca. 79,000 ha can be cultivated, a small area for a population of 540,000 (in 1981) heavily dependent on subsistence agriculture. The resultant population density, at least 6.8 persons per cultivated ha, is amongst the highest in the world, made more remarkable by the fact that most cultivated land consists of terraces on steep hillsides. "

"Cereals provide the main food component, especially rice and maize with smaller quantities of millet and wheat, but the districts appear to be doubtfully self-sufficient for food. Availability of food energy from food grown in the districts after allowing for payment of rents and taxes and for many sources of loss, probably scarcely exceeds the minimal subsistence need, said to be ca. 9 MJ per person per day. An earlier estimate of availability of ca. 5 MJ per person per day from local production was based on a cultivated area of only ca. 34,000 ha. (Mahat et al. 1987a). These data for the Project area are generally consistent with those for the Middle Hills given by Hrabovszky and Miyan (1987)."

"The provision of fodder for livestock is a complex operation with the proportions of various components varying with place and season of year. (In the two districts, livestock holdings by most households lie between two-thirds and one livestock unit per person.) Although much fodder is obtained from grass and trees growing on private agricultural land, especially uncultivated areas lying between the terraced arable land, fodder and bedding are the most significant forest products in terms of biomass used. Although animals are kept for milk production (buffaloes), draught power (oxen), or food (goats, etc), a major benefit of tethered and stall-fed animals, clearly perceived by the farmers, is obtained from their manure which is usually the only fertiliser applied to cultivated land."

"The farming practice is thus largely based on the transfer of nutrients from the forest soil, through leaves of trees and herbage, to farm animals and then to arable land in the form of manure and compost (Mahat et al. 1987a,b). The forest area to support this heavy use is of considerable extent, ca. 210,000 ha, but of this ca. 50,000 ha are in fact highly degraded shrubberies or grasslands, now almost devoid of trees, and large areas of the remainder are remote from habitation. Below about 2,000 m, forests (as relatively dense stands of trees) are usually small and scattered, though widespread.

The impact on the accessible forest is great and all forests near concentrations of households are suffering degradation as use exceeds growth (Mahat et al. 1987a)."

Forests in the Middle Hills play an important role in the subsistence economy and the multifarious linkages have been described and quantified by Mahat et al. (1987a,b). It is no exaggeration to say that the forests have underpinned the whole agricultural economy in the hills. They supply fuelwood for domestic and cottage industries use, fodder, manure (via animals eating forest fodder), wood for implements and construction, medicinal plants and the materials for traditional paper-making. Although substitution by non-forest products is possible in all cases, this situation seemed unlikely to be transformed on any scale in the foreseeable future, for socio-economic reasons. Already, much of the Middle Hills is a food-deficit area, importing food from the Terai. Economically, this imbalance has been supported by a slow diversification of income and employment in the Hills, even though subsistence agriculture remains predominant.

However, socio-economic change has been occurring in the Middle Hills at a pace and in directions that were unimaginable even in 1985. Studies conducted by Malla (1993) have documented the changes brought about in farming systems, especially the relationship between farm households and their surrounding forests, their livestock and their crops, due to the spread of the market economy and the creation of much more lucrative off-farm employment opportunities. Some agricultural land has been abandoned or converted (or allowed to revert naturally) to forest; livestock numbers have plummetted; chemical fertilisers are now readily available and subsidised by HMGN; many of the children who used to tend the livestock are now in school; and fodder requirements have declined. Fuelwood consumption has increased not for household subsistence use, but for small traders, tea houses and cottage industries. The number of trees grown on private land has increased substantially, not because of the eloquence of extension messages, but because the relative attractiveness or "profitability" has increased in comparison with other crops, land uses and household activities (Carter 1992). The presumption was that the new forests would supply fodder and fuelwood for subsistence, but increasingly it appears that commercial opportunities could become more important in the rapidly monetised rural economy.

Objectives of the Project

Phase 2 of the Project (NAFP/2: 1978-1985) began with tentative objectives and virtually no quantitative targets. This may now seem remarkable but realistically it could scarcely have been otherwise. The outline of the need of the Middle Hills could be indistinctly observed but there was no experience as to how that need could be met operationally. How much of what was needed? What species were best to meet the needs? Could they be grown successfully? How could local communities be brought into effective partnership? How could mutual trust be established? The questions were endless, the answers almost non-existent.

This very situation of almost unbounded ignorance, or at least uncertainty, determined the underlying philosophy of NAFP/2. Major aspects remained essentially unchanged in Phase 3, as follows:

1) No options should be closed off in the beginning. The Project was to assist in the totality of the forestry sector, with planting and protection on HMGN lands as well as on those of the communities. (Who would know where the boundaries would ultimately exist? Should not HMGN forestry be as much community oriented as "community forestry"? Would not confidence increase if local people had input to operations in HMGN forests?).
2) To begin with, the Project had to be small, so that all could learn together. By 1978, two panchayats stood out as foci of achievement and interest. There the Project began, and spread only as knowledge and confidence grew. Thus the number of nurseries grew from 5 in 1978, to 53 in 1983 and 117 in 1985. Correspondingly, the annual area planted or enriched increased from 100 ha in 1979, to 995 ha in 1983, to 1347 ha (at 112 different sites) in 1985, and to over 2000 ha in 1986, 1987 and 1988.
3) As expansion became possible, how would new panchayats for incorporation into project activities be chosen? By need, as determined by Project staff, or by local enthusiasm? The first option has never been chosen. As the possibilities for forestry action dawned upon a panchayat, usually by seeing what was going on in a neighboring panchayat, the leaders have approached Project staff, first for discussion, then seeking assistance. The initial moves for activity in a new panchayat therefore always arose with the panchayat, not the Project.
4) How can enthusiasm for and commitment to forestry be tested? Panchayats seeking assistance have in effect been asked four questions. Can you identify a site, acceptable to the Project for a nursery? Will you construct the terraces and other basic nursery component voluntarily, without pay? Can you identify sufficient areas of land for planting? Will you agree to control completely the grazing of animals and fodder collection so that new plantings

are not destroyed? The Project will provide no fencing. Will you cease to light fires in any area where an escape could damage the forest?

5) If a project is to succeed from a very uncertain knowledge base, it must have the operational mode of an experiment. Each step will provide new information which must be fed back into the forward planning of the next step of the project. With such a mode, the future in detail is always uncertain and may hold many surprises, often beneficial. Some examples help make the point.

Should the Project have consulted extensively to decide what proportion of which species should be grown to meet known or perceived deficiencies of forest products and then base its activities upon this information? After a very short time, this procedure was rejected because it was found that the most desired species were fodder trees. These proved to be very difficult to establish on denuded sites and even now, years later, the success rate is poor. Had a large part of the Project's efforts been devoted to solving this problem, as often advocated, little would have been accomplished, morale would have declined and impetus been lost. Instead, the option was taken of growing that which could be grown easily, even if the species were of lesser preference.

A native pine, *Pinus roxburghii*, proved particularly successful in establishing on denuded sites and soon the local people could see dramatic evidence of the success of their and the Project's efforts, as bare hills began to be covered in new forests. Pines are now being seen as more valuable by many farmers as long as they are not in extensive pure stands. Having planted pines to claim an area as "new forest", Project staff were surprised to find that at many sites a wide variety of broad-leaved species began to appear amongst the pines. The evidence was that the pines were acting as a psychological nurse crop (farmers did not allow grazing or browsing in the new forests and other uses were controlled) and as an ecological nurse crop (providing a less harsh environment for the growth of more sensitive species from old stumps, roots or seeds). All future Project operations and planning were changed to take maximum advantage of this occurrence. The problem of growing higher value species had been partly solved. Another aspect of the solution arose independently - growing the highest value species where they would be best protected and nurtured; on private land.

6) By agreement between HMGN and AIDAB, the Project is viewed as development assistance and not as a research project. This may seem to be at direct variance with what has been said above, but in reality it is not. We believe a great strength of NAFP has been its participatory action research mode of operation employed within a framework of appropriate flexibility approved by both HMGN and AIDAB.

7) During Phase III, the emphasis changed from establishing more nurseries and plantations, to the management of the forests planted 5-10 years earlier. Rather than conduct silvicultural trials to determine management options, Project staff tapped local experience and ideas arising from traditional and

indigenous practices. Almost every week, new examples of indigenous forest management systems were revealed and documented: almost everywhere that patches of trees existed (some less than 0.5 ha on a religious site, some over 160 ha) they were being managed by local people, in some cases with a history of continuous management going back more than 50 years. This recognition substantially altered the Project orientation and NACFP IV (1991-96) is designed to build upon and consolidate the management of all types of forests (natural and man-made) by User Groups.

8) Participatory Action Research requires continuous monitoring and periodic evaluation. Independent monitoring and evaluation has revealed how naive some of the 1978-85 assumptions were. Forests frequently transcend panchayat (now called Village Development Committee) boundaries, as do the User Groups. Such groups with clear mutually-recognised use rights may live in a different VDC to the forest that they manage. Some of the influential people who earlier requested plantations to be established on specific patches of land, were not the real users of the land or the forests. Some minorities were adversely affected by having their grazing lands converted to plantations, and so resisted and sometimes vandalised plantations that were "in the wrong place".

Project Management

The Project, strictly defined, has benefitted greatly from the Managing Agent being a university. A large amount of research, in the more traditional sense, has been undertaken by academic staff and students in association with, but not formally part of, the Project. The Project and the research work that has come to be associated with it have been mutually supportive and each has gained from the other. The managing agent has been the Australian National University, directly, or through its commercial arm ANUTECH Pty Ltd. In both cases, the University's Department of Forestry has provided the major professional, scientific and technical input.

Training at all levels, from worker to that of PhD, has been important. Large numbers of nursery foremen, senior foremen and other vocational level forest staff have been trained by the Project itself and these people have often become a most potent force for development. Without them, little could have been accomplished. With them, community forestry has gone a long way towards becoming an effective reality.

Neither professional nor vocational level staff can work effectively in an inappropriate institutional setting. Before 1977, almost no forestation had occurred in the Middle Hills. The main responsibility of staff was to control, whether by licensing or prosecution, the use of forest products and the institutional structure, and responsibilities of the Forestry Department at divisional (or later district) level reflected this. Inevitably, institutional features were therefore quite inappropriate and inadequate to support a major active programme of forestation, with a large technical component. The Project has therefore on the one hand necessitated, and on the other, assisted in, the beginnings of a gradual transformation of staff responsibilities and management and administrative methods to take account of the new commitments. The degree of change that has already occurred is indicated by the fact that the whole large-scale nursery and plantation establishment operation is now in Nepalese hands.

By the conclusion of NAFP/2 in 1985, much had been learnt about Project-community interaction and initiation of a successful plantation programme. A further 5-year phase of the Project (NAFP/3) commencing in 1986, placed the major emphasis on the devising and implementation of user level forest use and management agreements. The aim was to facilitate the transfer of responsibility from HMGN (and the Project) to the actual users of each individual forest area. This theme continues through Phase IV, with additional concern about the equitable utilisation of the newly-created plantations and the regenerated natural forests.

Benefit-Cost Analysis

General Approaches to Analysis

A good deal of confusion has been generated by diverse analyses of "the economics of Community Forestry in Nepal", few of which have precisely and rigorously defined the scope and perspective of the analysis. To clarify this situation, we attempt to classify several types of analysis that may be performed:

Type 1. A standard financial evaluation of the direct costs and returns to the owner(s) of the forest, (whether community, individual or government) relating primarily to wood production. Such analyses would include the direct costs and receipts, or equivalent market values where the products are used for subsistence. As an example, estimates of financial IRRs were derived in the Master Plan for the Forestry Sector of Nepal 1988 ranging from 13.3% for National Forests to 42.4% for a community where costs were offset by subsidies of various kinds.

Type 2. An analysis similar to Type 1, but inputs and outputs are valued at social or shadow prices. For example, the Nepal Master Plan reported an economic IRR of 25.4%, derived by adjusting **only** the wage rates in the above estimates, to reflect under- and un-employment. It is important to note that no adjustments were made for the social opportunity cost of capital, the shadow price of foreign exchange, and border pricing of wood outputs, nor was there any attempt to quantify any social, environmental or economic impacts of the Project, either on- or off-site.

Type 3. An analysis as for Type 2, but the returns to the households, and socio-economic effects **on-site** are included, e.g. improved agricultural yields due to windbreaks; reduced fuel collection times; improved nutrition to those directly involved; or the social cost of displacement of traditional users of the former grazing lands, who have not been included as forest beneficiaries.

Type 4. As for Type 3, but also including **off-site** "spillover" or "downstream" effects, e.g. socio-economic benefits resulting from the transfer of knowledge and skills generated by the Project, or physical environmental benefits such as reduced flooding or clearer water supplies. This could be considered a classical "complete" Benefit-Cost Analysis.

Type 5. This is a restricted form of benefit-cost analysis, which could be used, subject to careful interpretation, to evaluate the **Project, not community forestry** as such. The approach is based on the recognition that the benefits of R&D, training and extension programmes, socio-economic studies and baseline surveys cannot easily be quantified, even though ideally they should be included in the evaluation. A Type 5 analysis includes all "learning and servicing" costs but excludes the corresponding benefits. Such an analysis was conducted for the 1988 Mid-Term Review of NAFP/3 (AIDAB 1988), in which the estimated EIRR was 4%. The EIRR clearly would have been much higher if extra-production benefits had been included in the AIDAB analysis.

Selection of Evaluation Method

Type 1 analysis discussed above - financial evaluation - is quite straightforward. Once a particular management or technology package for forestry management has been specified, a financial analysis can be carried out. In practice, it has been financially viable for small-farmers and community groups to apply such knowledge for tree-farming, or to incorporate more trees into their farming systems, given present costs, expected yields, and market prices. Nevertheless, it is always difficult for the "foreign expert" to assess accurately the true shadow price of household labour (male, female, children, in different seasons etc) or the household's non-market evaluation of the worth of dung, fodder, various types of fuel, or other by-products: conceptually straightforward, but almost impossible in practice. Byron (1991) was forced to the conclusion that if a farmers' informed and rational behaviour on whether or not to grow trees contradicted the recommendations from his economic models, then either the data in the model or the decision criteria are probably incorrect.

Type 2 analysis is unsatisfactory as a socio-economic Benefit-Cost Analysis. Its scope is far too narrow.

Type 4 is theoretically the ideal form of analysis, but is too broad to permit practical assesssment. Attempts to measure demonstration benefits of the Project have been limited (Hamilton 1985), and there is little point in conducting a detailed economic analysis of the off-site physical effects, as the Project has been assumed to make a negligible difference.

Type 3 analysis would thus appear to be most practicable form of economic appraisal for the NAFP, at least in the first instance. Projects based on reforestation will succeed only if:

i) the Government of Nepal perceives overall economic benefit in such projects and

ii) people living adjacent to forested areas perceive sufficient financial benefit to themselves to participate in establishment and protection of the forestry resource.

In conducting a Type 3 analysis, attention can be focused on the direct onsite benefits and costs, and expenditures on ancillary support and research excluded. Human Resource Development expenditures can be kept separate from the Afforestation investment, as a "grant" which does not have to pass the test of a BCA investment appraisal. It is still difficult at this stage, nevertheless, to carry out a Type 3 economic evaluation of the Project. There is little available information on appropriate border prices, and markets for wood products in Nepal are in a controlled and vestigial state. A financial analysis can be undertaken, but this is only indicative of a proper economic appraisal.

Subsequently, it might be possible to conduct a Type 5 analysis. NAFP has large R&D and training/extension components, as well as serving as a pilot for numerous other forestation projects in Nepal and elsewhere. If all costs of R&D and demonstration are included in the analysis, then ideally the corresponding outputs or returns should be quantified and valued, under the "matching principle". Yet it is very difficult to assess the value to Nepal of more or better-trained forest rangers, or more foresters with post-graduate degrees from Australia, sociological research, recommendations for legal, institutional and land-tenure reforms, etc.

It can be argued that, under a Type 5 analysis, it would be unnecessary to quantify **all** the benefits and solve for an EIRR, if the EIRR were found to be "acceptable" regardless of R&D and demonstration benefits. Alternatively, the threshold value can be calculated that must be imputed for R&D and demonstration benefits to make the EIRR "satisfactory".

Financial Analysis for Local Households

Local Impacts

In overall terms, the Project is considered a profitable investment for Nepal. To be a success it must be of direct benefit to local communities. Worldwide, there appears to be relatively poor correlation between national economic welfare and benefit to local people in the design of many forestry projects, even "community forestry" projects (Douglas 1983).

In the case of NAFP, there is convincing evidence that the distribution mechanism and tenure aspects are working well from the viewpoint of local communities. Local community participation in project activities is high, and wilful damage to established forests is low. Although forest watchers hired from local communities are necessary, fencing is not, so the resource protection costs in the project are relatively low.

With respect to local communities, the Project has had a number of important spinoffs, affecting private tree growing.

Private Forestry

The direct impact of the Project on attitudes to, and progress towards, re-establishment of tree cover by owners on private land in the region is difficult to quantify. One thing is clearly established: in the period 1964-1988, spanning the main period of activity of the Project to date, tree cover on private land in the two Districts has increased significantly.

An increase in tree cover was evident in a Project study by Carter and Gilmour (1989) where tree cover on a transect across the two Districts in 1964 was estimated from aerial photographs and compared with cover measured in a field transect in 1988. Extrapolating the results for all privately owned upland terrace areas in the two Districts, the authors deduced that tree cover increased by an equivalent of 6500 ha. Substantial increases in tree cover have occurred also on non-cultivated private land - an equivalent of 16,000 hectares between the two measurement years.

It cannot be assumed that this change in tree cover is wholly or even primarily a result of Project activity. Carter and Gilmour point out that increases have occurred elsewhere in Nepal outside the range of influence of the Project or similar projects. The results support the view that forest area in the Middle Hills has changed little in recent decades.

The Project may be having some impact on the rate at which trees are being established in the region, and possibly on the quality of resource. Apart from providing better access to good quality seedling stock through its nursery component than would otherwise have existed, the Project also provides important technical and demonstration benefits. The Project has shown that, through the indirect mechanism of pine establishment, local broadleaved species valued for fodder, litter and other purposes can be re-established, where direct attempts to grow them on cleared areas had failed. Now, the Project is achieving considerable progress with re-establishment of forest cover on degraded shrubland areas. There is a great deal of this type of land in the region. It is highly probable that, currently and potentially, the Project can strongly influence the rate at which reforestation will occur. However, as mentioned earlier, there is a complex interplay of rapidly changing factors affecting household decisions about crops, livestock and trees - about the balance between subsistence and cash-cropping, even the decision about whether to continue farming at all.

Evaluation of Welfare Effects for Households

The above discussion suggests that the Project is favourably affecting the economic welfare of most local inhabitants, and is recognised by them as doing so. It could be argued that a typical farm budget or discounted cash flow analysis could be conducted to "prove" to farmers that it is financially worthwhile for them to include some tree-growing in their farming systems. However, this appears to be unnecessary, for two reasons.

First, farmers have already performed such calculations subjectively, have made their assessments, and have started planting trees in many places. In certain locations where farm families have chosen not to plant trees, privately or communally, there is strong evidence to suggest that this was a rational decision based on a correct assessment, given their highly-localised supply-demand situation for forest outputs, and other options and constraints.

Second, given that farming is undertaken largely by subsistence households, it is virtually impossible to second guess or "shadow price" **all** of the non-priced inputs and outputs for households from the forest - for example, women and child labour, leisure time and interactions with agriculture. When there is no bank credit, paid labour force or commercial product, the notion of a "farm budget" is severely limited.

On-Site Returns and Costs of Plantation Forests

In contrast to tree-growing by subsistence households, consider a plantation investment on government or community land using paid employees - the type of NAFP/3-assisted forestation activity which is amenable to Discounted Cash Flow or Benefit-Cost Analysis. Relevant calculations are shown in Table 7.1 for a period of 24 years. These estimates form the basis of an economic evaluation of NAFP/3.

Revenue Estimates

Production. Gilmour and Shepherd (1988) have provided figures on biomass output figures for forests in the Project area, as shown in Table 7.2. The estimates are supported by field measurements to year 12. Beyond that the figures are estimates.

Prices of Wood. No freely functioning markets for wood exist (officially) in Nepal. A ban on sales of wood from private land remained in force until 1988, and has been relaxed and re-imposed since. Most wood entered the market via government corporations (abolished in 1990) whose pricing policies were insulated from market forces. The market prices used in the analysis below are not strictly financial, but represent the potential financial prices that could be attained at that time.

Sawlogs. Data on prices for sawlogs are limited. Some price estimates are available from a USAID study (Shaikh et al. 1988) in the Kathmandu Valley. They range from Rs 70 to Rs 220 per cft. Higher prices than these can apply. Shaikh et al. noted an average price for 10 foot 3" x 4" beams of Rs 270/cft, equivalent to Rs 9500/cu m. Large profit margins appear to be accruing to retailers at these prices, and if normal profits prevailed throughout, standing timber prices would potentially be higher. Using the volume-weight conversion factors in Commonwealth Forestry Handbook, of a standing price for medium grade structural timber, a residual stumpage price can be estimated at Rs 959/cu m or Rs 2131/tonne.

On this basis, use of any figure below Rs 1000/cu m (i.e. Rs 2220/tonne) for medium grade structural wood on the stump would seem unduly pessimistic.

Table 7.1.
Proposed Planting and Value of Production

Year	Plantation Forest (ha)	Protected Forest (ha)	Output Plant. Forest '000Rs	Output Prot. Forest '000Rs	Total Output '000Rs	Total Costs '000Rs	Net Revenue '000Rs
(1)	(2)	(3)	(4)	(5)	(6)	(7)	(8)
1	2,086	500	135	44	179	25,369	-25,189
2	1,940	500	261	222	484	23,429	-22,944
3	1,500	500	353	536	890	26,830	-25,939
4	1,500	500	446	978	1,424	25,390	-23,965
5	1,500	500	535	1,467	2,002	26,065	-24,062
6	1,250	500	1,472	2,101	3,573	9,891	-6,317
7	1,250	500	2,339	2,763	5,103	9,755	-4,651
8	1,250	500	5,145	3,460	8,606	10,219	-1,612
9	1,250	500	6,430	4,192	10,623	10,683	-59
10	1,250	500	7,907	4,959	12,886	11,146	1,719
11	1,000	0	18,441	5,781	24,222	11,544	12,678
12	1,000	0	28,284	6,488	34,772	11,342	23,429
13	1,000	0	35,982	7,107	43,090	11,740	31,350
14	1,000	0	43,883	7,598	51,482	12,137	39,344
15	1,000	0	51,717	8,070	59,788	12,535	47,253
16	750	0	58,352	8,417	66,769	12,866	53,903
17	750	0	64,982	8,735	73,718	12,598	61,119
18	750	0	71,360	9,019	80,379	12,929	67,449
19	750	0	77,896	9,296	87,193	13,260	73,932
20	750	0	84,369	9,558	93,927	13,592	80,335
21	750	0	89,668	9,741	99,409	13,923	85,486
22	750	0	94,963	9,906	104,869	14,254	90,614
23	750	0	100,004	10,042	110,046	14,585	95,460
24	750	0	105,205	10,178	115,383	14,917	100,466

This assumes that all structural wood in the Project area is potentially saleable in Kathmandu and that the owners and controllers of the resource wish to sell it in that market. Whether the material could be brought to market is a function of community choice, and the transport logistics. An approximate calculation, made with Project staff assistance in Kathmandu, suggests that all the resource in the project area could be carried to roads in range of Kathmandu for only a few days' return walk. A guideline price for sawlogs in the project area, calculated on the basis of potential marketability in the Kathmandu Valley, should be set at no less than Rs 1750/cu m, with a more likely price being Rs 2000/cu m.

Fuelwood. Official royalty prices for fuelwood (split wood from logs) in Nepal range from Rs 50/tonne to Rs 100/tonne. To determine the potential financial price of fuelwood, such figures are of little value. Gilmour and Shepherd's production figures (see Table 7.2) classify fuelwood into two categories: stemwood, which is a highly valued product sold in markets for specific uses in the Monsoon season; and branchwood, much of which tends to be collected and sold in less formal ways. Bazaar prices have exceeded Rs 750/tonne for stemwood. Hamilton (1985) estimated an opportunity cost for collection of fuelwood, but this does not help to formulate a financial price on the stump for branchwood.

If stemwood fuelwood were harvested and sold in freely operating markets by the owners of wood in the Project area, it might be worth Rs 600-800/tonne at the stump. It is more difficult to estimate an intrinsic value on the stump for branchwood, given the lack of formal markets, the inadequacy of the opportunity cost methodology for valuing on the stump, and the difficulty of using substitution price data. Given rapid increases in values for wood in Nepal, a figure of Rs 200/tonne on the stump for this material would seem more reasonable as a potential financial price than the present figure based on an administered royalty of Rs 100/tonne. If Government prices for fuelwood were to be applied to private supply as well, it is unlikely much would be formally traded, due to the opportunity cost to the private sector. The implicit price is thus higher.

Gilmour and Shepherd (1988) estimate that about 30% of total fuelwood production is stemwood; the remainder is branchwood. Using these proportions and the above prices gives a weighted average stumpage price figure for fuelwood of Rs 320 to Rs 380 per tonne.

Fodder/litter Prices. A recent review of NAFP by the Australian International Development Assistance Bureau (1988) cites various earlier studies which have attempted to value litter and fodder material, based on opportunity costs. From these studies, the AIDAB study postulated a basic fodder value of Rs 250/tonne, and a litter value of Rs 100/tonne. These estimates were adopted in the analysis.

Prices Used in Analysis. The prices for forest products assumed for the economic analysis are as follows:

Table 7.2.

Average Biomass Output per Planted Hectare (tonnes)

Year	Fodder	Fuelwood	Timber	Litter
1	0.26	0.00	0.00	0.00
2	0.26	0.00	0.00	0.00
3	0.25	0.00	0.00	0.00
4	0.25	0.00	0.00	0.00
5	0.24	0.00	0.00	0.00
6	0.53	0.90	0.00	0.00
7	0.51	0.90	0.00	0.00
8	1.89	2.60	0.00	0.25
9	0.47	1.80	0.00	0.25
10	0.65	2.30	0.00	0.50
11	1.03	1.94	2.16	0.50
12	1.02	1.94	2.16	0.75
13	1.01	1.94	2.16	0.75
14	1.00	1.94	2.16	1.00
15	1.00	1.94	2.16	1.00
16	1.00	1.94	2.16	1.00
17	1.00	1.94	2.16	1.00
18	1.00	1.94	2.16	1.00
19	1.00	1.94	2.16	1.00
20	1.00	1.94	2.16	1.00
21	1.00	1.94	2.16	1.00
22	1.00	1.94	2.16	1.00
23	1.00	1.94	2.10	1.00
24	1.00	1.94	2.16	1.00
25	1.00	1.94	2.16	1.00
26	1.00	1.94	2.16	1.00
27	1.00	1.94	2.16	1.00
28	1.00	1.94	2.16	1.00
29	1.00	1.94	2.16	1.00
30	1.00	1.94	2.16	1.00
31	0.64	0.38	0.00	0.52
32	0.64	0.38	0.00	0.52

Table 7.2 Continued

33	0.65	0.38	0.00	0.52
34	0.65	0.38	0.00	0.52
35	0.65	13.00	24.90	0.52
36	0.66	0.38	0.38	0.52
37	0.66	0.38	0.38	0.52
38	0.67	0.38	0.38	0.52
39	0.67	0.38	0.00	0.52
40	0.68	33.40	65.30	0.52
41+	0.70	1.80	3.00	0.60

Source: Gilmour and Shepherd (1988)

sawlogs: Rs 1750/tonne and 2000/tonne
fuelwood: Rs 320/tonne and 380/tonne
fodder: Rs 250/tonne
litter: Rs 100/tonne

Cost Estimates

The project costs for NAFP in 1986 and 1987 were Rs 25,369,000 and Rs 23,429,000 respectively. These figures include all Nepal Government Forestry Department costs in the two Districts. Attention must be paid to the proportion of Nepal Government costs related directly to plantation activity under the Project, as opposed to what the Government would have spent on forest supervision in the area without the Project. Very little plantation would have resulted from the regular programme of Government forestry expenditure. In this case, an adjustment was made by subtracting the regular programme expenditure (Rs 4,575,775 per annum) from the Project budget, so that only incremental costs are incorporated.

In Table 7.1, costs of plantings are shown for 24 years, reflecting the expectation that plantation establishment by the Forestry Department and local communities would continue beyond 1990, the terminating year of NAFP/3 (and even 1995 and NACFP IV).

Results

Using the estimates in Table 7.1, and assuming a sawlog price of Rs 2000/tonne and a fuelwood price of Rs 380/tonne, the IRR for NAFP/3 was calculated at 13%.

Various forms of sensitivity analysis were conducted:

- Lowering the price of sawlogs to Rs 1750/tonne and the fuelwood price to Rs 320/tonne, giving an IRR of 11.5%.
- Reducing all yields by 10% which lowers the IRR from 13.0% to 12.0%.
- Increasing costs by 10% which reduces the IRR from 13.0% to 11.9%.
- Calculating the IRR for the Project assuming no further plantings after 1990, giving an IRR of 10%.

Varying the discount rate had a significant effect on the estimated NPV. The NPV of the Project ranged from zero with a discount rate of 13% to Rs 228.6 million with a discount rate of 5%. This is commonly the case with forestry projects.

Off-site Aspects of the Project

There are two senses in which a project such as NAFP might have economic or environmental effects outside the two Districts in which it operates. The first is the demonstration effect of the project: the dissemination throughout Nepal of the ideas, technological developments, and social and policy mechanisms successfully applied in the Project to implementing community forestry.

The second is the physical, downstream impact that a project concerned with radical changes to land use might potentially have, especially in a steep and unstable terrain.

Demonstration or Spillover Effects

There is no way at present of quantifying the extent to which technology and training inputs from the Project will induce plantation areas or productivity to increase in the rest of Nepal. All that can be done, under the circumstances, is to demonstrate the impact on returns to the Project of assumed levels of external benefits of this process.

Such a calculation moves beyond financial aspects to economic evaluation of the project, and should be considered only in the context of a full economic appraisal. However, the sensitivity of the financial analysis to inclusion of spillover effects is of interest, and the calculation is made for this reason.

Hamilton (1985) carried out a calculation by assuming that NAFP would boost production in other plantation projects outside the two Districts by 10% as a minimum and 20% as a maximum. He assumed that such external benefits would take some time (two years) to appear in other projects and that after 10 years the demonstration effect would be exhausted.

Hamilton also assumed that plantation projects outside the two Districts would establish a total area of 5000 ha per annum on average. This estimate is conservative, given the large World Bank Community Forestry Project and other international inputs to the forestry sector.

If Hamilton's assumptions are applied (using Project cost and price data as described above) to calculate external "spillover" benefits attributable to NAFP, the following conclusions can be drawn. Applying a 10% spillover effect to the revenue stream shown in Table 7.1, the IRR would increase from 13% to 15.1%. The 20% spillover assumption takes the IRR to 16.9%. A threshold valuation was not calculated, as the Project was considered acceptable even without imputing the demonstration or spillover values.

Physical Downstream Impacts

As already explained, the natural geological processes of erosion and sedimentation in Nepal far exceed those attributable to the activities of man. Any downstream benefits of the project, except on a very small scale adjacent to Project operations, will be concealed by natural events. No physical downstream benefits are therefore claimed for the Project.

It must be emphasised that this surprising rejection of benefits is peculiar to the geologic and climatic environment of the Middle Hills of Nepal. It may not be applicable in other countries or geomorphological circumstances.

Overall Economic Appraisal

A full economic evaluation of the Project, using appropriate border prices, shadow pricing of factor inputs and so on has not been possible in this analysis. General experience with forestry projects in developing countries suggests that economic returns will usually exceed financial returns, largely because economic prices for output generally exceed financial prices. That would probably also apply in this case, especially when spillover effects in agriculture and forestry are taken into account. It is important to restate that the costs and prices used in the above analysis are really a mixture of financial and economic values.

The Project's main effects will be upon communities in the Project area, as was the intention of the Project. Downstream benefits cannot be expected from the Project. Taken in isolation, no major impact on urban or extra-regional supply of wood products will arise. The Project is likely to have significant spillover or demonstration effects, especially important in reversing forest degradation presently occurring in Nepal. However, quantifying these effects and assigning them to the Project is at present a highly speculative exercise. The calculations performed for spillover effects suggest that economic returns to the project are likely to exceed significantly the imputed financial returns.

Sensitivity of the financial analysis to cost and yield variations is not high, but the Project was shown to be sensitive to the discount rate. The response depends to some extent on the length of the evaluation period, but more importantly on what occurs in the Project area after the 5-year project period.

The major uncertainty in a project of this nature is probably one already accounted for in sensitivity testing: the possibility that, after cessation of specific project activities, no further planting would be undertaken. Returns to the Project are surprisingly robust even under the assumption of no further planting after the initial 5-year period; the IRR is still 10%.

Biological uncertainties, such as late-developing adverse growth patterns, the advent of pests and pathogens, and so on, should be addressed in the evaluations.

Given the strong probability that economic returns to a project of this nature will exceed the calculated financial returns, and given the relative robustness of the project in relation to various types of risk and uncertainty, it can be concluded that the Project is potentially of significant economic benefit to Nepal.

Conclusions

General Findings

There are several general findings from the above analysis which bear on policy issues for forestry in Nepal.

The role of man in massive erosion and downstream sedimentation in Nepal must be kept in perspective. In volume terms across whole catchments, erosion and sedimentation are very largely attributable to natural geological causes, and human action is most unlikely to control or even significantly modify these large scale events. The most significant consequence of human action in these matters is the loss of top soil from over-utilised or poorly utilised areas. Here amelioration can occur, ably assisted by reforestation.

A shortage of forest products is presently threatening the quasi-subsistence agriculture of the Middle Hills. Deforestation has occurred over many centuries often driven by forces which scarcely exist today. The present distribution between forest and non-forested land in the hills was largely created before rapid population expansion occurred in the middle of this century. Recent change has therefore largely been a degradation of forests within relatively stable boundaries rather than deforestation by conversion to agriculture.

Successful community-scale forestry will often require large changes to legislation and in bureaucratic organisation. Here, Nepal has been exceptionally fortunate because over past decades legislative change has been closely coordinated with new understandings. The newly published Master Plan for Forestry and the latest amendments to the Forest Act are in accord with the developing thrust of community-based forestry.

Projects such as NAFP can be justified in Nepal on the basis of immediate wood, fodder and litter benefits to local communities, enhanced control over localised environmental degradation, and important spin-off effects in agriculture and grazing. At this stage of reforestation in Nepal, NAFP also has important spillover benefits due to its role in innovating workable community participation models for forestry, and in solving biological and technical problems within the constraints of available human and technological resources.

Policy Implications

Various policy implications flow from these and other observations raised in this study.

Reforestation in Nepal needs to be predicated upon, and justified by reference to, the physical and economic benefits which are clearly achievable. These do not include significant mitigation of the massive cycle of erosion and sedimentation taking place in areas of the sub-continent, appealing as such a benefit may be in justifying projects to donor agencies and the international community in general.

Technological problems remain in the establishment of forests in the Hill country of Nepal. However, these should not be an isolated or even the major thrust of project work.

Implementation of reforestation policies which, in Nepal, implies the development of workable community based methods, is the major objective. Projects should at all stages take account of, and as far as possible incorporate into project design and implementation, local knowledge, organisations and aspirations. They should be designed to deliver immediate and obvious benefits to those people most directly involved in local land use. In most cases, this will involve changing perceptions and methodologies within Government, at least as much as within the local communities themselves.

Socioeconomic analysis will need to pay much greater attention than in the past to the likelihood of a project attaining over time such a character that the local communities will see it as "their own". Only then will long-term sustainability be assured. Rarely, however, will such an ideal state be attained in the short term because of inevitable errors in project formulation, design and implementation. All participants will need to learn from their joint, and sometimes initially misunderstood, experiences but aim steadily at the target of essentially local effective management of a new or recreated resource.

References

Australian International Development Assistance Bureau (1988) 'Mid-Term (Draft) Review: Nepal Australia Forestry Project, Phase 3', AIDAB, Canberra.

Belbase, N. and B.W.Boer (1993) 'Nepal Country Report', in *Yearbook of International Environmental Law 1992*, Graham and Trotman Ltd, London.

Bremmer, C.N. (1987) 'The Role of Vegetation and Land Use in Flooding, Erosion and Wasting in the Ganges Drainage Basin, India: A Literature Review', Department of Hydrogeology and Geographical Hydrology, Free University, Amsterdam.

Byers, A.D. (1987a) 'An Assessment of Landscape Change in the Khumbu Region of Nepal Using Repeat Photography', *Mountain Research and Development*, 7:77-81.

Byers, A.D. (1987b) 'Landscape Change and Man-Accelerated Soil Loss: The Case of the Sagarmatha (Mt Everest) National Park, Khumbu, Nepal', *Mountain Research and Development*, 7:209-216.

Byron, R.N. (1991) 'Cost-Benefit Analysis and Community Forestry Projects', In D.A. Gilmour and R.J. Fisher (eds.) *Villagers, Forests and Foresters: The Philosophy, Process and Practice of Community Forestry in Nepal*, Sahayogi Press, Kathmandu.

Carson, B. (1985) 'Erosion and Sedimentation Processes in the Nepalese Himalaya', ICIMOD Occasional Paper 1, Kathmandu.

Carter, A.S. and D.A. Gilmour (1989) 'The Dynamic Nature of Trees or Private Farm Land in Central Nepal', *Mountain Research and Development*, 9:381-391.

Carter, E.J. (1992) 'Tree Cultivation on Private Land in Nepal's Middle Hills: An Investigation into Local Knowledge and Local Needs', OFI Occasional Papers, No 40, Oxford University.

Douglas J.J. (1983) *A Re-appraisal of Forestry Development in Developing Countries*, Martinus Nijhoff Publishers, Dordrecht.

Eckholm, E.P. (1975) 'The Other Energy Crisis: Fuelwood', Worldwatch Paper 1, Worldwatch Institute, Washington, DC.

Fisher, R.J. (1989) 'Institutional Incompatibility in Community Forestry: The Case of Nepal', in J. Fox and R.J. Fisher (eds) Community Organisations and Government

Bureaucracies in Social Forestry, Working Paper No. 22, Environment and Policy Institute, East-West Center, Honolulu.

Gilmour, D.A. (1986) 'Reforestation or Afforestation of Open Land: A Nepal Perspective', in A.J. Pearce and L.S. Hamilton (eds) *Land Use, Watersheds and Planning in the Asia-Pacific Region*, FAO Regional Office for Asia and the Pacific, Bangkok, pp. 157-169.

Gilmour, D.A., M. Bonell and D.S. Cassells (1987) 'The Effects of Forestation on Soil Hydraulic Properties in the Middle Hills of Nepal: A Preliminary Assessment', *Mountain Research and Development*, 7: 239-249.

Gilmour, D.A. and K. Shepherd (1988) Unpublished mimeo, NAFP, Kathmandu.

Gilmour, D.A. (1985) 'Reforestation or Afforestation of Open Land: A Nepal Perspective' paper presented at the International Workshop on Landuse Planning in a Watershed Context, Gympie, Queensland.

Gilmour, D.A. and R.J. Fisher (1991) *Villagers, Forests and Foresters: The Philosophy, Process and Practice of Community Forestry in Nepal*, Sahayogi Press, Kathmandu.

Goldstein, M., J.L. Ross and S. Schuler (1983) 'From a Mountain Rural to a Plains Urban Society: Implications of the 1981 Nepal Census', *Mountain Research and Development*, 3:61-64.

Griffin, D.M. (1988) 'Innocents Abroad in the Forests of Nepal', ANUTECH Australian National University, Canberra.

Griffin, D.M., K.R. Shepherd and T.B.S. Mahat (1988) 'Human Impact on Some Forests of the Middle Hills of Nepal: Comparisons, Concepts and Some Policy Implications', *Mountain Research and Development*, 8:43-52.

Hamilton, C. (1985) 'Benefit Cost Analysis', Appendix B in Project Document, *Nepal Australia Forestry Project - Phase 3*, Australian International Development Assistance Agency, Canberra.

Hamilton, L.S. (1986) 'Towards Clarifying the Appropriate Mandate in Forestry for Watershed Rehabilitation and Management', Reprint 94, East-West Center Environment and Policy Institute, Honolulu.

Hamilton, L.S. (1987) 'What are the Impacts of Himalayan Deforestation on the Ganges-Brahmaputra Lowlands and Delta: Assumptions and Facts', *Mountain Research and Development*, 7:256-263.

Hrabovszky, J.P., and K. Miyan (1987) 'Population Growth and Land Use in Nepal', *Mountain Research and Development*, 7:264-270.

Ives, J.D. (1987) 'The Theory of Himalayan Environmental Degradation: Its Validity and Application Challenged by Recent Research', *Mountain Research and Development*, 7: 189-199.

Ives, J.D. and B. Messerli (1989) *The Himalayan Dilemma: Reconciling Development and Conservation*, United Nations University and Routledge, London.

Mahat, T.B.S. (1985) Human Impact on Forests in the Middle Hills of Nepal, PhD dissertation, Australian National University, Canberra.

Mahat, T.B.S., D.M. Griffin and K.R. Shepherd (1987a) 'Human Impacts on Some Forests of the Middle Hills of Nepal: Forests in the Subsistence Economy of Sinhu Palchok and Kabhre Palanchok', *Mountain Research and Development*, 7:53-70.

Mahat, T.B.S., D.M. Griffin and K.R. Shepherd (1987b) 'Human Impact on Some Forests of the Middle Hills of Nepal: A Detailed Study in SE Sindhu Palchok and NE Kabhre Palanchok', *Mountain Research and Development*, 7:111-134.

Mahat, T.B.S., D.M. Griffin and K.R. Shepherd (1986a) 'Human Impact on Some Forests in the Middle Hills of Nepal: Forestry in the Context of the Traditional Resources of the State', *Mountain Research and Development*, 6:223-232.

Mahat, T.B.S., D.M. Griffin and K.R. Shepherd (1986b) 'Human Impact on Some Forests of the Middle Hills of Nepal: Some Major Human Impact Before 1950 on the Forests of Sinhu Palchok and Kabhre Palanchok', *Mountain Research and Development*, 6:325-334.

Mahat, T.B.S., D.M. Griffin and K.R Shepherd (1985) 'An Historical Perspective of the Forests of Sindhu Palchok and Kabhre Districts of Nepal', *Schweizerishe Zeitschrif für Forstwesen*, Suppl. 74:235-244.

Malla, Y.B. (1993) The Changing Role of the Forest Resource in the Hills of Nepal, PhD dissertation, Department of Forestry, Australian National University, Canberra.

Myers, N. (1986) 'Environmental Repercussions of Deforestation in the Himalayas', *Journal of World Forest Resource Management*, 2:63-72.

NCS (1988) *Building on Success: The National Conservation Strategy for Nepal* HMG Nepal/World Conservation Union, Kathmandu.

NPC and IUCN (1992) *A Legislative and Institutional Framework for Environmental Management in Nepal,* HMG Nepal/World Conservation Union, Kathmandu.

Pandey, A.N., P.C. Pathak and J.S. Singh (1983) 'Water, Sediment and Nutrient Movement in Forested and Non-Forested Catchments in Kumaun Himalaya', *Forest Ecology and Management*, 2:19-29.

Pearce, D., E. Barbier and A. Markandya (1990) *Sustainable Development: Economics and the Environment in the Third World*, Earthscan, London.

Sattaur, O. (1987) 'Who is Moving the Mountains of Nepal?', *New Scientist*, 1544:29.

Shaikh, A., K. Kanel, D. Koirala, S. Pandey and P. Thopa (1988) 'Some Preliminary Summary Data Tables form the Draft Report on Forest Products Marketing in the Kathmandu Valley', unpublished mimeo.

Shrestha, R.L. and D.B. Evans (1984) 'The Private Profitability of Livestock in a Nepalese Hill Farming Community', *Agricultural Administration*, 16.

Shrestha, C.L. and B.K. Uprety (1988) 'Environmental Concern Leading to Formulation of Impact Assessment: Nepal', paper presented at Expert Group Meeting on Environmental Impact Assessment of Development Projects, ESCAP, Bangkok.

World Bank (1978) Nepal Forestry Sector Review, No 1952-NEP.

Chapter 8

THE PALAWAN REGIONAL DEVELOPMENT PLAN

GEOFF GOLDRICK AND DAVID JAMES

This case study demonstrates the use of benefit-cost analysis in broad-scale regional development planning in the Philippines. The relevant area is the Island of Palawan.

The case study is an excellent example of pro-active planning and assessment, with a strong emphasis on protection of the natural resource base. Maintenance of sustained yields from natural resources of the island is a strong focus of the study.

The chapter begins with a review of development plans and strategies for environmental protection in Palawan. An explanation is provided of the planning and assessment methodology. Details are then presented of the overall regional development plan. To demonstrate the application of economic valuation techniques at a more local scale, the study concludes with an analysis of environmentally interconnected activities in Bacuit Bay on the northeast coast of Palawan: logging, tourism and fisheries.

Background

Palawan Island, the principal island of the Province of Palawan, lies to the west of the central Philippines and north of Sabah (see Figure 8.1). As a result of its relative isolation and virulent malaria, Palawan has until recently been underpopulated and undeveloped in comparison with the rest of the Philippines. Rapid population growth, caused by high fertility rates and in-migration from

D. James, The Application of Economic Techniques in Environmental Impact Assessment, 217–246.

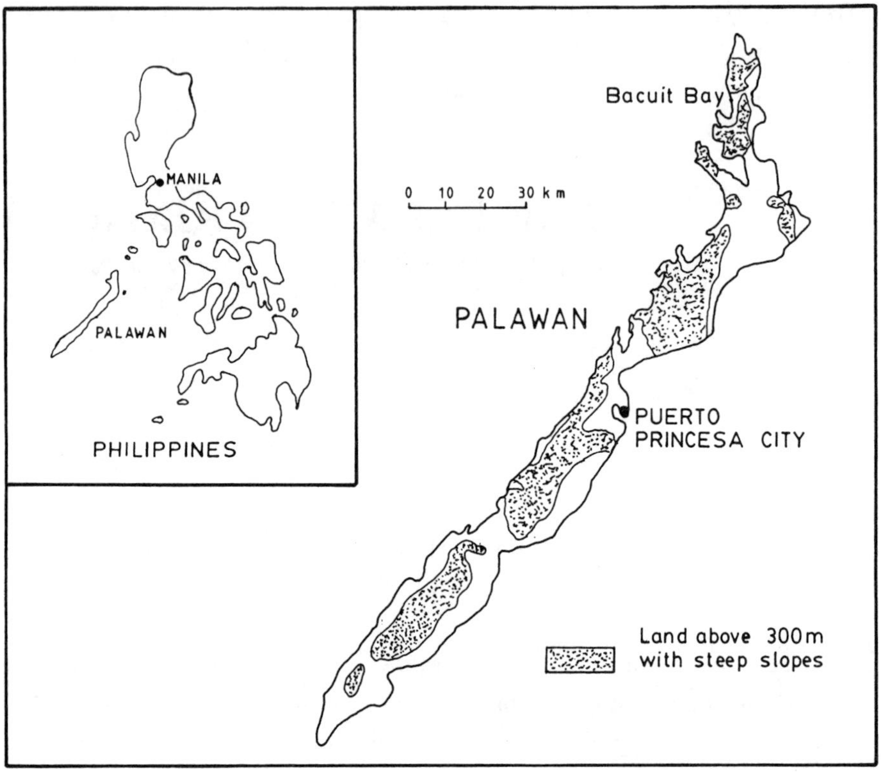

Figure 8.1. Map of Palawan.

over-populated areas of the Philippines, has recently begun to threaten the island's fragile and largely unspoilt environment.

In 1979 the Philippine Government's National Council on Integrated Area Development initiated the Palawan Integrated Area Development Project (PIADP). The purpose of the project was to promote carefully managed development of the island. An integrated multisectoral approach was adopted. A key component of PIADP was the Integrated Environmental Program (IEP), designed to assess the environmental impacts of various development strategies, involving various mixes of exploitation and conservation, on a regional scale. The main assessment tool was benefit-cost analysis. Favoured strategies were then combined to produce a Strategic Environmental Plan (SEP).

The more local study of Bacuit Bay was undertaken separately by researchers at the East-West Center (Hodgson and Dixon 1988). As an exercise in economic analysis, it is a fitting complement to the PIADP, and exemplifies the kind of detailed planning that can be expected for individual watersheds and coastal areas, once development gets under way.

Description and Uses of Palawan's Natural Resources

Physical Environment

The island of Palawan, located in the south west of the Philippines, is flanked by the South China Sea to the west and by the Sulu Sea to the east (see Figure 8.1). It is 425 km long and varies between 5 and 40 km in width. The island is predominantly mountainous, its steep, narrow spine rising to a high point of 2,085 m. It has a tropical monsoonal climate characterised by high rainfall intensities.

Palawan's flora includes substantial areas of pristine vegetation, hence Palawan has been termed "the last ecological frontier" of the Philippines. Dipterocarp and scrub forests blanket the steep slopes, intercepting rainfall and reducing the potential for rapid run-off and soil erosion. Palawan contains the last remaining stand of casuarina forest in the Philippines. Palawan is noted for its mangrove stands. Approximately 46,000 ha of mangrove form a discontinuous belt along the coastline. The entire island has been declared a Mangrove Swamp Forest Reserve. Mangroves are, however, being increasingly exploited for timber and the making of charcoal.

A number of rare and endemic species of fauna can be found in the island's forests, including the peacock pheasant, which is listed in the IUCN's Red Data Book. The island was declared a Game Refuge and Bird Sanctuary in 1967. However, the trade in native wildlife continues.

The major land use in Palawan is forestry. Agriculture is the next most important activity, with 8% of the land area under permanent agriculture. The remaining agricultural area is cropped under a form of shifting cultivation known as "kaingin". Kaingin has been responsible for much of the environmental degradation occurring in the montane areas of the island. Palawan has important marine resources. The island's coastal waters supply 66% of the total fish catch of the Philippines. Illegal fishing methods and overfishing threaten the long term future of this resource. Abundant coral reefs provide the habitat for numerous commercial fish species, and constitute an underexploited resource for the development of tourism. Other tourist attractions include the island's majestic and diverse scenery, an underground river, karst landscape and pristine beaches.

Palawan's mineral wealth is thought to be considerable. Only a limited amount of mining and quarrying activity exists at present. There appears to be little recognition of the environmental despoilation that these activities can create, and environmental protection measures are lacking.

Social and Demographic Trends

In 1983 the population of Palawan was 318,000 resulting in an average density of 27 people per sq km. This contrasts with the average for the whole of the Philippines of nearly 170 people per sq km. A high rate of immigration from other islands is expected to result in a trebling of the population by the year 2007. The majority of Palawans live in rural areas. Urban dwellers presently constitute only 24% of the population, although this proportion is increasing.

Role of Forests

In 1968 forest cover accounted for approximately 92% of the land area, but by 1980-83 it had been reduced to 68-70% (780,000 ha). This substantial loss is attributable to uncontrolled shifting cultivation and damaging logging practices. Rates of forest loss at the time of the study were approximately 19,000 ha/yr.

Palawan's forests perform a vital role in stabilising the island's ecosystem. In areas where the forest cover has been disrupted, increased run-off strips the topsoil and facilitates landslides and mud flows. Nutrients are washed away and, in the absence of forest cover, are not replaced. Vegetation is slow to re-establish and runoff is drastically increased. The combined effects are lower and more erratic river base flows, and more frequent and severe flooding downstream. Large cobbles and much silt is transported to lowland areas.

Flooding has caused damage to property and crops, and deposition of coarse sediments has lowered the agricultural value of lowland flood plains. Sediments carried from the hillsides have blanketed the sea bed, damaging or killing sea grasses and coral reefs, and reducing fish numbers and diversity. Irrigation schemes have been threatened by the reduction in river base flows.

Threats to the Natural Environment

Palawan's population has been growing at 4.6% per annum compared with the Philippines national average of 2.7%. There are fears that this rapid population growth could stress the local environment beyond its self-sustaining capacity if development were not properly managed. Environmentally damaging practices such as open cast mining, kaingin and unsustainable logging have been intensifying.

The economic benefits of such activities are essentially short term. In the long term, large losses of economic welfare can be expected. Such losses are associated mainly with negative externalities affecting other industries which offer the prospect of sustainable development, notably fishing, agriculture and tourism. The ability of Palawan to support its projected population in the medium to long term will depend on the sensible development of these sustainable industries, despite the possibility of some temporary reductions in income.

It has been recognized that development must respect the environmental limits set by the tropical climate, the unusual topography and the sensitive ecology of the island, otherwise environmental degradation will continue to accelerate, reducing economic prospects in the medium and long term. The problems predicted from uncontrolled development include (PIADP 1985):

- further spread and intensification of kaingin, overexploitative logging, and hence loss of forest;
- loss of native forest flora and fauna as forest retreats;
- erosion of unprotected catchments with a consequent increase in flooding, sedimentation, destruction of roads and irrigation systems, reduction of economic potential of fishing and tourism industries by damaging fish life and destroying coral;
- pollution of the environment by unregulated mining activities, agriculture, urban and industrial activities, illegal fishing methods and hunting of rare fauna;
- encouragement of tourism for short-term gains, leading to the loss of natural assets.

Ultimately the combined impact of such activities would be a contraction of the economic base and an inability to support the increasing population of the island.

Environmental Policy and Legislation

(Ben Boer and Angus Martyn)

The Philippines has a well developed environmental policy, set by Presidential Decree in the 1970s. Section 4 of the *Philippine Environmental Policy P.D.* No 1151 states:

"All government agencies and private corporations should prepare an environmental impact statement for every project undertaken which would significantly affect the quality of the environment. Such document should present in detail the following information:

a) the environmental impacts of the project;
b) unavoidable environmental effects of the project;
c) alternatives to the proposed action;
d) measures that shall be undertaken so that the short term uses of resources of the environment will be consistent with the maintenance of enhancement of the long-term productivity of the same;
e) whenever a proposal involves the use of depletable or non-renewable resources, a finding must be made that such use and commitment are warranted."

This section forms the basis for the Environmental Impact Assessment system established by *Presidential Decree P.D. No* 1568 of 1978. P.D. 1586 commenced its effective operation in July 1982. Commenting on the Decrees, del Rosario (1988) stated:

"P.D. No 1151 and 1586 highlight environmental impact assessment (EIA) as a planning tool for environmentally sound development (sustainable development) which means an assurance that the maximum utilisation of resources for human purpose should be conducted in the most economically efficient manner and in a way that creates the least adverse impacts to the biogeophysical and social environment. EIA can therefore be defined as a process which project proponents have to conduct and which consists of identifying and predicting the impacts of projects and programmes on the environment and in interpreting/communicating information of such impacts in a manner which can be utilised by planners and decision-makers to effect sustainable development."

Since 1987, the EIA system has been run by the Environmental Management Bureau (EMB). Section 16 of the *Executive Order* No 192 established the Bureau within the Department of Environmental and Natural Resources (DENR). A number of groups and individuals within the EMB have particular responsibilities for different aspects of the EIA system. These are (Brillantes 1992):

1) Environmental Impact Assessment Group (EIAG). The EIAG carries the principal responsibility for implementing P.D. 1586. It functions as a central point for the submission of major assessments, undertakes preliminary reviews of these, coordinates involvement of other government agencies, makes recommendations concerning the environmental acceptability of the

project, and monitors the proponent's compliance with conditions set under the assessment. It also undertakes general administration and reform of the EIA system.

2) Environmental Impact Assessment Review Committee (EIARC). A multi-disciplinary committee, the membership of the EIARC is drawn from the private sector, academe, relevant consultants or government agencies outside the EMB. EIARC acts as an "expert" group in reviewing the adequacy of the Environmental Impact Statement (EIS) submitted by the proponent. Its conclusions are of an advisory nature only.

3) EMB Director. The director forwards the recommendations of the EIAG and/or the EIARC to the Undersecretary for Research and Environment in DENR. The Undersecretary, having the power to decide on whether a project is approved or not, is the ultimate decisionmaker under the EIA system.

Away from the centrally based EMB, the DENR maintains small EIS Units in each of its regional offices. As well as having responsibilities for site inspections and compliance monitoring, these Units perform an evaluation and administrative role for projects requiring less complex forms of EIA.

There also exist Environmental Units which are, in effect, small teams designated in the lead agencies to provide technical assistance to a range of parties. Lead agencies are those government departments or authorities with responsibilities for particular types of developments of projects. These Environmental Units are meant to coordinate with the EMB to achieve integration of national environment policy, including EIA.

Environmental Impact Assessment

Under the Presidential Proclamation 2146 of 1981, EIA is required for two major categories of projects: environmentally critical projects (ECPs) and projects in environmentally critical areas (ECAs). This proclamation, and some subsequent refinements of it, were designed to limit the number of projects subject to EIA.

Environmentally critical projects include:

- heavy industry (steel, oil, gas and petroleum and petro-chemicals);
- resource extraction (mining, forestry, fishery, mangrove products, grazing);
- infrastructure projects (major dams, power plants, reclamation project roads and bridges).

Environmentally critical areas include:

- national parks and reserves;

- areas set aside for aesthetic potential for tourism;
- habitats of endangered flora and fauna;
- areas of unique historic, archaeological or scientific interest;
- areas which are traditionally occupied by cultural communities or tribes, or contain cultural monuments;
- areas prone to natural calamities;
- areas with critical slopes;
- areas classified as prime agricultural lands;
- water resource areas;
- fisheries;
- certain mangrove areas;
- coral reefs.

If the project appears to be an ECP, the proponent is required to submit sufficient information to the EMB to determine if in fact this is the case. A full EIS is required for ECPs.

If the project is to be carried out in an ECA, a project description (PD) must be submitted to the appropriate DENR regional EIS Unit unless exempted. Exemptions apply if the discharge of waste is minimal and easily managed, the project has a capitalisation value of under P 500,000 and employs less than 20 persons. After review of the PD, the Unit makes a judgement whether the possible environmental effects of the project warrant the preparation of an EIS.

The following steps (in summary form) are involved in the Philippine EIA system (adapted from Quisumbing and Brillantes 1992):

Environmentally Critical Projects

1) Project Description prepared and submitted.
2) Status as an Environmentally Critical Project confirmed.
3) Environmental Impact Statement prepared and submitted.
4) Preliminary Review of EIS by EIAG may recommend any of the following:
   - issuance of an Environmental Clearance Certificate (ECC),
   - requirement of confirmation of information contained in the EIS,
   - referral to EIARC for further evaluation.
5) The EIARC may require that a public hearing be held if the project may affect a large number of people or if the project generates significant public interest.
6) After review of the EIS, the EIARC may recommend:
   - the issue of an ECC,
   - denial of an ECC,

- that additional information be obtained.

7) After all review procedures have been completed, the EIAG forwards its recommendations to the Undersecretary for Research and Environment in DENR.
8) The Undersecretary makes the final decision whether to issue an ECC. The project may be implemented if it is granted.

Projects in Environmentally Critical Areas

1) Determination whether project is exempted from the EIA process.
2) If not exempted, Project Description prepared and submitted.
3) Review and Evaluation by DENR regional EIS Unit. The Unit may recommend any of the following:
   - issuance or denial of an ECC,
   - submission of an EIS,
   - submission of further information.
4) Should an EIS be required, the EMB may become involved and the process becomes similar to that for steps (3)-(8) for ECPs as shown above.
5) If an EIS is not required, the EIS Unit forwards its recommendation to the DENR regional Executive Director. The Executive Director is delegated to make the final decision whether to issue an ECC. The project may be implemented if an ECC is granted.

The EMB has the power to suspend or cancel an ECC if the EIA requirements or the terms or conditions of the ECC are breached. In addition, or as the alternative, a breach can attract a fine of up to 50,000 pesos.

The Philippines Environmental Policy specifies that the EIS must contain a detailed statement of the environmental impacts of the proposal, any adverse effects which cannot be avoided if the proposal is implemented, any alternatives to the proposed action, a determination that the short term uses of the resources are consistent with the long term productivity of those resources and, where the resources are non-renewable, a finding must be made that such use and commitment are warranted.

The level of technical compliance with EIA system has increased substantially in the last five years (Brillantes 1992). However, substantial problems are still being experienced in screening projects sufficiently so that the volume of assessments does not exceed the institutional capacity of the EMB and the regional offices. The EIAG itself consists of only five non-technical and six technical staff, of whom only one member has had a formal education in EIA (Brillantes 1992). As a consequence, it is probable that the quality of EIA remains inconsistent and time taken for processing of applications uncertain.

This section draws heavily upon the work of Amy Brillantes, Environmental Manager, BHP Engineering Philippines, Manila. Her cooperation is gratefully acknowledged.

Integrated Planning for Palawan

The Palawan Integrated Area Development Project (PIADP)

The need for a coordinated plan leading to sustainable development resulted in the creation of the Palawan Integrated Area Development Project (PIADP) by the Philippines Government in 1979. Its aims were the initiation of integrated economic development whilst safeguarding the island's natural heritage. PIADP was the fifth integrated area development project to be undertaken by the Philippines Government's National Council on Integrated Area Development (NACIAD), but the first to include specifically an environmental component.

The Integrated Environmental Program (IEP)

The environmental component of the PIADP, the Integrated Environment Program (IEP), was in effect a regional environmental development project. The IEP was conducted in two phases. Phase I ran from February 1983 to October 1983, with Phase II beginning in September 1984. Funding for the programme was obtained from the European Economic Community.

The primary objectives of IEP (Phase I) were:

- to assess the existing environmental situation in Palawan;
- to recommend measures for consideration and inclusion in the detailed planning and implementation of other PIADP components;
- to identify short and medium term actions to counter adverse development impacts, prevent further degradation and rehabilitate damaged areas;
- to recommend a long-term environmental regulation plan for mainland Palawan.

The objectives for Phase II were:

- formulation and setting up of an environmental monitoring and evaluation system;
- undertaking a specific study of the Tamlang catchment, leading to the formulation of a rehabilitation action plan for the Tamlang and associated catchments;
- preparation of a Strategic Environmental Plan, with details of its translation into an implementation plan.

The Strategic Environmental Plan (SEP)

Over an 8-month period in 1983-84, detailed investigations were conducted of Palawan's natural resources, resulting in the completion of the SEP in 1985. The SEP was designed to provide a comprehensive framework for the sustainable development of the island of Palawan. It consisted of a series of integrated projects and programmes operating over a period of ten years. Long term responsibility for the implementation of the Plan was vested in the Palawan Environment Staff (to be created during phase II).

As specified by Hunting/MacDonald/PMDSI (1985) the objectives of the SEP were:

- to stabilise the environment within a ten year period;
- to develop Palawan's natural resources to the fullest possible extent, consistent with sound long term management concepts;
- to conserve Palawan's unique or special landscapes and wildlife with benefits to science, tourism and recreation;
- to accommodate the expected increase in population in ways consistent with maintaining a stable environment;
- to create optimum conditions for health, employment and income generation;
- to establish a sound socio-economic balance between resource exploitation and environmental protection;
- to define short and medium term measures for environmental protection and rehabilitation, especially in environmentally sensitive parts of the island;
- to provide guidelines for effective long-term regulatory environmental control;
- to integrate the implementation of short-, medium- and long-term planning through the coordinated participation of all relevant agencies.

The programmes and policies making up the Plan fall under several groups of strategies: zonal; suprazonal; social and institutional; and research, extension and education. Zonal strategies are those in which land uses, such as forestry and agriculture, are regulated to conform with the conditions of the ecological zone to which they are specific. Suprazonal strategies relate to activities that are not restricted to a particular ecological zone or subzone, eg road construction, mining and tourism. Social strategies cover specific social issue such as land tenure, and upland settlement.

The major elements of the Plan are described below.

1) Measures to stabilise the environment, with particular reference to:
   - protection of the forest on the steeper uplands;

- better regulation of commercial logging;
- improved and accelerated land classification of forest lands to improve regulation or control of settlement within them;
- protective management of coastal mangroves;
- measures to introduce stable systems of agriculture in already occupied areas, including increased emphasis on the present Upland Stabilization Program (USP);
- special studies and remedial programmes for endangered catchments, especially in the southeast;
- final selection and establishment of National Parks and Reserves, both terrestrial and marine, to protect Palawan's unique wildlife and encourage tourism.

2) Development of activities aimed at improving or maximising settlement and production in environmentally sound locations so as to attract settlement away from environmentally vulnerable areas. Such activities include development of roads, water supplies and education and health facilities. Special emphasis would be placed on:
    - developing stable farming systems in the low hills zone;
    - intensifying agriculture and settlement on the lowland zone;
    - accelerated land classification and titling in both these zones;
    - developing the full potential for irrigation.
3) Developing a mechanism for plan implementation involving:
    - setting up the Provincial Environmental Staff (PES) and the Environmental Monitoring and Evaluation System (EMES);
    - strengthening key agencies involved in plan implementation, especially the Bureau of Forest Development (BFD), and those agencies responsible for marine resources development and regulation;
    - developing local awareness of environmental problems.

The structure of the SEP is shown in Figure 8.2 with proposed programmes and projects.

Integration of Environment and Development

Palawan's growing population and dependence on its natural resource base has necessitated the integration of environmental protection and economic development. Without environmental protection, economic activity would destroy the resource base on which it depends. Without development, the environment would be degraded as a result of poverty, and induced mismanagement and overexploitation of resources. The SEP aimed to demonstrate the mutual benefits by examining the economic costs and benefits of environmental protection strategies.

Provisions for Individual EIAs

The plan contains provisions for environmental impact assessment (EIA) of development activities, such as mining, transport infrastructure, and urban and industrial growth, which have the potential for significant site-specific environmental impacts.

EIAs should be undertaken as and when resource use conflicts arise. Economic analysis can be used as the evaluation methodology for EIAs, and can also assist the design of optimal mitigation measures. An example of the way in which economic analysis can be used to resolve resource conflicts is described in the last section of this case study.

Economic Evaluation of Environmental Components of the IEP

Approach to Economic Analysis

Economic analysis of the Palawan Integrated Environmental Plan was undertaken by consultants to the programme and reported by Finney and Western (1986). The SEP was seen principally as a strategy for environmental protection with a strong development element, not as a development strategy with a strong environmental element. The high value of returns from development activities in the plan, for example from the establishment of stable farming, threatened to overshadow returns from environmental activities. A decision was made to exclude development benefits and focus on the benefits of environmental protection.

This separation of analysis in one sense contradicts the basic philosophy of benefit-cost analysis. If an integrated plan is proposed then the benefits of that plan should be considered in total and compared with a baseline situation where there is no planning. Alternative strategies should be assessed in a similar manner. Nevertheless, the evaluation does indicate the economic value of resource protection measures.

The main benefits of the plan were identified as:

- ensuring an adequate standard of living for the island's growing population;
- maintaining sustained forest production;
- maintaining sustained fisheries production;

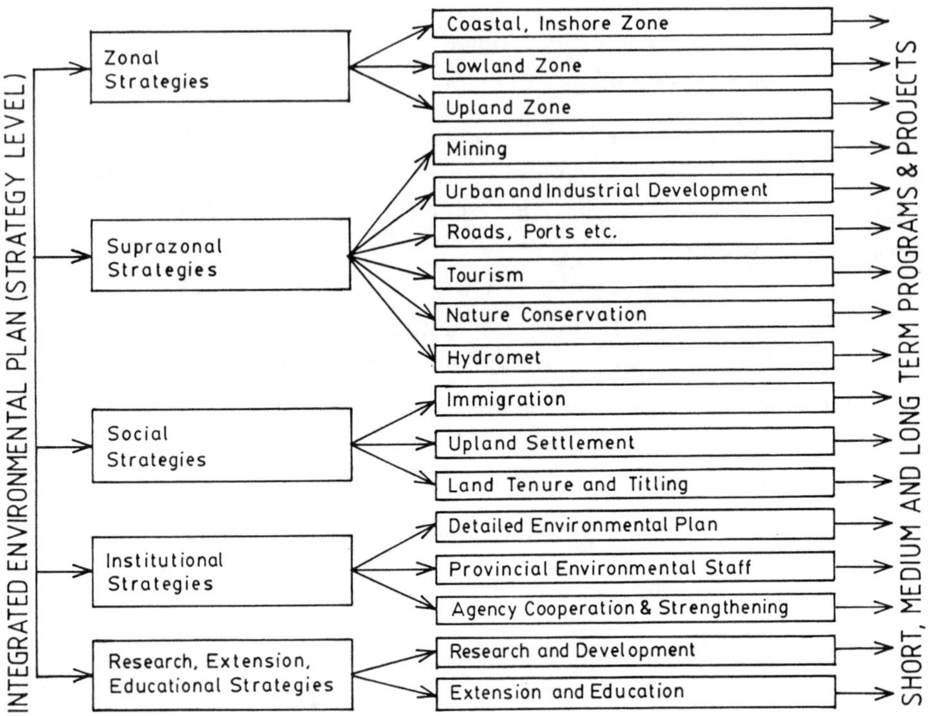

Figure 8.2. Structure of the Strategic Environmental Plan for Palawan.
Source: Derived from Hunting/PMDSI/MacDonald (1985).

- growth in tourism;
- catchment control to reduce the risks of flood damage.

The costs associated with the plan were considered mainly in financial terms, ie the costs of resources devoted to environmental protection and the short term losses resulting from constraints on development. The economic analysis included only the quantifiable benefits of the plan. Non-quantifiable social and ecological benefits were excluded.

For assessment purposes the components of the plan were divided into three main groupings:

- land based protection;
- mangrove conservation;
- inshore marine conservation.

A separate economic analysis of each component was not considered feasible because various benefits were generated by two or more components. Instead, the following approach was adopted:

1) The capital and recurrent costs of each component were estimated.
2) Benefits were divided into "separable" and "common" sets. Separable benefits are those that can be attributed to a particular component whereas common benefits are those which can be attributed to two or more components.
3) The value of each type of benefit was estimated for each group of components.
4) An assessment was made of separable costs and benefits for mangrove protection and land-based protection. This analysis was intended to judge whether either of these two protection groups was economically justified even if common benefits were disregarded. Inshore marine conservation could not be assessed in this way because all of its benefits were considered to be common.
5) Finally, the costs and benefits of each of the components were combined to give an overall benefit-cost comparison for the environmental protection components of the Integrated Environmental Plan.

In order to calculate the NPV of the different strategies, discount rates of zero and 5% were applied. These rates were chosen because they were considered to reflect the importance of long term costs and benefits. The time period for the analysis was 100 years. This unusually long period was chosen because many of the economic benefits of environmental protection would accrue only in the distant future. At any positive discount rate, benefits and costs occurring more than 50 years in the future would have fairly insignificant present values.

Environmental Protection Benefits

To estimate the benefits of land-based environmental protection, mangrove protection and inshore marine conservation, land-based protection and mangrove protection were divided into separable and common categories. The separable benefits of land-based environmental protection were taken to include:

- improved agricultural returns, especially irrigated rice;
- sustained forestry production;
- reduction of flood damage to infrastructure.

The common benefits included:

- fisheries production;
- tourism benefits.

The benefits of land-based environmental protection were valued using the productivity approach and the replacement and repair cost approach. The benefits accruing to agriculture result from the stabilisation of hydrological regimes due to the protection of river catchments. The main benefit would be the prevention of large scale losses of present and planned irrigated rice land as a result of changes in water availability. In addition, the loss of rainfed crop areas as a result of erosion or siltation by flood sediments would be avoided. These benefits could be valued using the productivity approach. A cost to agriculture would be the loss of production (estimated to equal 2.2 million Pesos) resulting from the substitution of forestry for kaingin in catchment areas.

The forestry production benefits result from the introduction of sustainable yield practices in the forested areas. The benefits would take the form of timber production in commercial logging concession areas, the collection of minor forest products, and future timber production in reafforested areas. For the purposes of the economic analysis the second two types of benefit were ignored as they were considered negligible relative to the overall costs and benefits. The commercial forestry returns were calculated for the with- and without- environmental plan situations. With no control over commercial logging, harvesting rates would exceed sustainable yields. The plan involved a reduction in forestry income in the short and medium term but a higher sustained income in the long term. An important consideration is that the relative price of tropical hardwood was likely to be higher in the future, thus magnifying projected benefits under the plan.

The protection of upland catchment areas would result in a reduction in potential flood damage to infrastructure. The benefits of protection can be valued in terms of the repair and replacement costs of damaged infrastructure that are avoided. These costs were estimated to be about 9 million Pesos per annum. This calculation ignores the costs of traffic disruption.

Catchment protection would reduce the siltation of inshore waters and thus protect sensitive coral reefs and fish breeding grounds. As fisheries benefits were common benefits these were not quantified separately for land-based protection measures. Benefits from tourism would be another common benefit, shared with inshore marine conservation.

Mangrove Conservation Benefits

Mangroves offer an effective and natural form of coastal protection. Acting as sediment traps, they help protect coral reefs from river sediments. Mangroves provide a useful source of timber and charcoal as well as providing a nursery ground and habitat for commercially important fish species.

In the absence of the environmental plan it was estimated that one third of the island's mangroves would be destroyed in the next few decades. The harvesting of mangroves would result in short term gains that would diminish substantially over the years. With the plan, net returns from mangrove forests would remain at a steady 36.7 million Pesos per annum, ignoring the benefits of mangroves for marine life.

Benefits of Inshore Marine Conservation

The inshore marine area, with its productive coral reefs, mangroves and sea grass beds, is a nationally important fisheries resource. Control of illegal fishing and ensuring the survival of the reef would be required to maintain the long term productivity of the resource. A thriving reef community is one of the major tourist attractions of the island.

The benefits of inshore marine conservation could not be evaluated separately from the rest of the plan, as they are also attributable in part to mangrove conservation and land-based protection measures.

Environmental Protection Costs

Three main groups of environmental protection costs were examined. These were marine conservation costs, land-based and mangrove protection costs, and other environmental protection costs. Land-based and mangrove protection costs were grouped together because both types of environmental protection would be undertaken primarily by the Bureau of Forest Development (BFD). The land-based and mangrove protection costs are described below.

Strengthening of the BFD was considered essential to establish effective control over forest resources. Costs of the BFD strengthening programme were estimated at 2.31 million Pesos for capital costs and 2.65 million Pesos for annual recurrent costs. Expansion of the BFD was due to take place regardless. For this reason, the consultants assumed that only two thirds of the strengthening costs could be charged to the environmental plan.

For reforestation and forest rehabilitation in degraded catchments, a capital cost of 1000 Pesos per ha was assumed. An area of 40,800 ha was in need of reforestation and rehabilitation. Total capital costs were calculated as 40.8 million Pesos. The capital costs of forest preservation were also assessed. Forest preservation consists of creating forest reserves where only limited exploitation of minor forest products is allowed. The only significant costs of forest preservation other than those already covered by the BFD strengthening programme were the costs of reserve demarcation. At an assumed cost of 1500 Pesos per 1000 hectares, the total cost of demarcating 331,000 ha of forest was estimated at 496,000 Pesos.

Marine conservation would require the strengthening of the Bureau of Fisheries and Aquatic Resources (BFAR), involving an expansion of staff, vehicles and boats, office facilities, and equipment. This was calculated to cost 2.15 million Pesos in capital expenditure and an increase in recurrent expenditures of 2.42 million Pesos per annum. As the greater share of this expenditure was targetted towards fisheries development, as opposed to pure marine conservation, only one third of the costs of strengthening the BFAR were included in the analysis. In addition to strengthening the BFAR, marine conservation was to be achieved by the establishment of three marine parks at an estimated cost of 0.25 million Pesos.

Although much of the responsibility for direct environmental protection lies with the BFD and BFAR, five other agencies involved directly or indirectly with Palawan's physical environment would require environmental protection staff. On the basis of one Environmental Officer for each of the five organizations, the recurrent costs were estimated at 150,000 Pesos per annum. The capital costs of creating these posts were considered negligible and were therefore ignored.

The total capital cost for the environmental protection components of the Integrated Environmental Plan amounted to 43.81 million Pesos, most of which would go towards reafforestation and forest rehabilatation. Total operating costs were 3.24 million Pesos, of which 80% would be allocated to strengthening of the BFD and BFAR.

All costs attributed primarily to development were omitted. The costs of establishing specific environmental institutions required to implement the plan were also omitted. These costs belonged to Phase II of the Integrated Environmental Program. They were not considered high enough to affect the economic analysis of the plan.

Economic Evaluation of Environmental Protection Components

A benefit-cost analysis was undertaken of the land-based environmental protection measures, comparing their costs with the separable benefits directly attributable to land-based protection. As outlined above, the cost items of land-based protection consisted of the strengthening of the BFD, reforestation and rehabilitation, forest preservation and loss of kaingin crop production in the mountains zone. The separable benefits included increased potential for irrigated agriculture, sustained output in the forestry sector, and reduction in flood damage to infrastructure.

The NPV was calculated over a 100 year time horizon at discount rates of zero and 5%. The results were:

NPV at 0% = 26,332 million Pesos

NPV at 5% = 714 million Pesos

Internal rate of return = 6.1%.

It must be remembered that this analysis did not include the common benefits of tourism and inshore marine fisheries. Nor does the analysis include many ecological and social benefits.

The direct cost of mangrove conservation was calculated as the proportion of the strengthening of the BFD that could be attributed to this task. One-third was considered a reasonable estimate. Capital costs were therefore 0.5 million Pesos over two years and recurrent costs were 0.6 million Pesos per annum. The assumption was made that, without protection, 14,600 ha of mangrove would be destroyed by 2007. The economic results were as follows:

NPV at 0% = 2,373 million Pesos

NPV at 5% = 150 million Pesos

Internal rate of return = 8.7 per cent.

The significant benefits likely to accrue for inshore marine fisheries were not included in the analysis.

General Economic Evaluation of the Strategic Environmental Plan

The benefit-cost analysis of the environmental plan included all separable and common benefits of all the environmental protection components of the IEP as well as the costs of mangrove protection, land-based protection and inshore marine conservation. The benefits of inshore marine fisheries and tourism were added to the separable benefits of land-based environmental protection and mangrove protection. Table 8.1 shows a breakdown of costs and benefits according to the year in which they accrue. The overall results were:

NPV at 0% = 33,356 million Pesos

NPV at 5% = 1,384 million Pesos

Internal rate of return = 7.0%.

The estimates quoted above represent a lower bound estimate of the economic worth of the integrated environmental plan because the direct benefits of development components were omitted. Social and ecological benefits were also excluded.

Ecological benefits of the plan would be difficult to value. For example, the implementation of the plan should prevent the loss of endemic and other more widely distributed species. It is difficult to place a value on the maintenance or loss of genetic diversity, as in most cases the genetic bank is still a potential resource to be developed at some future date. Wild animals and plants also have a value beyond their use value.

Some social benefits could be treated in a similar way to ecological benefits. By preventing environmental degradation the IEP may, for example, ensure that fishermen retain their livelihoods and avoid the social and psychological disruption and hardship that might occur if the fisheries were to decline. Unquantifiable benefits such as these should always be presented in qualitative form alongside the quantitative results of the economic analysis. Ignoring unquantifiable benefits can result in a substantial underestimation of the benefits of a plan.

Resource Management in Bacuit Bay

Bacuit Bay as a Local-Scale Study

In 1985 logging operations were commenced in the upland catchment area of Bacuit Bay. These operations resulted in sedimentation and damage to coral reefs and marine fisheries in the Bay, adversely affecting the viability of the two other important industries of the area, fishing and tourism. Hodgson and Dixon (1988) undertook a study in which they examined the economic effect of logging in Bacuit Bay.

Table 8.1.

Economic Costs and Benefits of Palawan's Integrated Environment Plan.
(Pesos million/annum at 1983 prices)

Year	Land-based Protection	Mangrove[a] Conservation	Marine Conservation		Tourism	
	Net Cash Flow	Net Cash Flow	Benefits	Costs	Benefits	Net[b] Cash Flow
(1)	(2)	(3)	(4)	(5)	(6)	(7)
1	-1	-16	-	1	-	-18
2	-201	-16	3	2	-	-216
3	-195	-15	6	2	-	-206
4	-191	-15	9	2	1	-198
5	-184	-14	12	2	1	-185
6	-175	-14	15	2	1	-173
7	-183	-13	18	2	1	-179
8	-174	-13	21	2	2	-166
9	-157	-12	24	2	2	-145
10	-149	-12	28	2	2	-133
11	-140	-11	31	2	2	-120
12	-131	-10	35	2	2	-106
13	-124	-10	38	2	3	-97
14	-114	-9	41	2	3	-81
15	-107	-8	44	2	3	-70
16	-98	-8	47	2	3	-58
17	-89	-7	50	2	4	-44
18	-81	-7	53	2	4	-33
19	-72	-6	56	2	4	-20
20	-64	-6	58	2	4	-10
21	-56	-5	61	2	5	3
22	-50	-5	64	2	5	12
23	-42	-4	67	2	5	24
24-100	376	12	70	2	6	462

Notes: [a] Including only the timber and charcoal benefits, not the fisheries.

[b] This net cash flow is calculated by adding together the land-based protection and mangrove conservation net cash flows and the marine conservation and tourism benefits and then deducting the marine conservation costs.

Source: Finney and Western (1986)

Description of Bacuit Bay Area and its Major Economic Activities

Bacuit Bay is situated in the northern tip of Palawan and covers an area of 120 sq km (see Figure 8.3). The Bay contains five islands with nine outer islands on its seaward shelf. The islands are surrounded by coral reefs. Coral reefs also form a band along the bay's coastline, interrupted only where river breaks occur. Three small rivers and several seasonal streams enter the bay. The Manlag River is the largest, receiving 70% of runoff from the Bacuit Bay catchment. The catchment covers 78 sq km of land rising up to the central Palawan dividing range. Land use in the drainage basin consists primarily of primary and secondary forest.

Logging operations in the Bacuit Bay drainage basin occur as part of a larger logging concession which extends to the north and south including most of the remaining primary forest in northern Palawan. Commercial and artisanal fishing is practised all year round in Bacuit Bay and on its seaward shelf.

Tourism is little developed in Palawan because of its remote location and fears of malaria infestation, and has only recently begun to play a role in the Bacuit Bay area. In 1982 a scuba diving resort catering mainly to Japanese tourists was established on one of the islands close to the entrance of Bacuit Bay. The success of this resort prompted the opening of another diving resort on an adjacent island.

Options for Forestry Management

The economic analysis undertaken by Dixon and Hodgson focused on the present value of the gross revenues of the three industries in the area under two different scenarios:

Option 1 - Logging is banned in the Bacuit Bay drainage basin. This option allows the tourism and fisheries industries to continue without being subjected to additional damage from logging-induced sedimentation.

Option 2 - Logging proceeds without intervention in the Bacuit Bay drainage basin. This option maximises revenue to the logging concessionaire but results in a loss of revenue to the two other industries.

Assessment Methodology

The first stage of the assessment was to predict the physical environmental impacts of the two options. A study of the inter-connected terrestrial and marine ecosystems of the area was undertaken and used to predict the environmental impacts likely to accompany each of the options over a ten year period. The

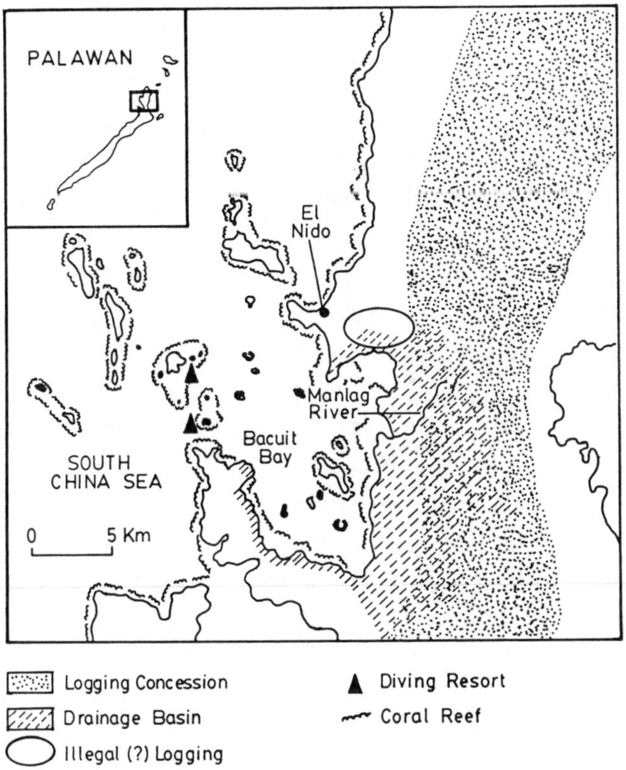

Figure 8.3. Study area of Bacuit Bay.
Source: Hodgson and Dixon (1988)

environmental effects of the two options were then quantified. Likely changes in the gross revenues of the three industries with and without a logging ban were estimated for this ten year period.

The present value of gross revenue accruing to each industry over a period of ten years was calculated and compared for the two options. Discount rates of 10% and 15% were used, reflecting the rates used in the Philippines in both private and public project evaluation. Such rates are also fairly standard for multilateral and bilateral aid agency appraisals.

Environmental Impacts of Forestry Management Options

The impacts of logging on catchments include damage to young trees as a result of unplanned felling, soil compaction, and wind and water induced erosion. The major cause of increased erosion in logging sites is the construction of road and skid-trail networks. Erosion induced by logging increases sediment loads in rivers and streams, detrimentally affecting fish and invertebrate biomass and diversity, and damaging the coastal marine environment.

The environmental impact of Option 2 was predicted over a ten year time horizon by extrapolating the known effects of logging in the initial year of 1985. These ongoing physical effects were measured in the field by Hodgson in 1986. Effects included accelerated soil erosion, high levels of suspended sediment load, sediment discharge into Bacuit Bay, reductions in diversity and biomass of reef fauna including coral, and sediment deposition in Bacuit Bay.

Measurements were made of sheet erosion. Results from experimental plots showed sheet erosion to be 240 times greater in plots in logged areas than in plots located in primary forest. The contribution of other forms of erosion, such as earth and mud slides, rill wash and gullying, were not evaluated. Their contribution is likely to be highly significant, as evidence of all three processes was abundant. This was confirmed by sedimentation monitoring which showed the total sediment load to be more than twice the load attributable to sheet erosion. It was noted that recolonisation of eroded areas by plants was slow, often allowing erosion to continue for one or two years after cessation of operations.

Of the sediment reaching the mouth of the river, a proportion is deposited in the bay, with the remainder transported further into the ocean. The distribution of the particles is dependent upon many factors, including their size, the density of the water, and the prevailing wind. Sediment deposition throughout the bay was measured using sediment traps.

From the data obtained, the total tonnage of sediment deposited in the bay was calculated for the different seasons. A control station provided an estimate of the amount of sedimentation expected in the absence of logging. This figure was subtracted from the figures recorded in Bacuit Bay. Coastal erosion was not thought to contribute to deposition, as the coastline of Bacuit Bay is relatively stable. Re-suspension by currents may have been substantial. The figures obtained must be regarded as rough approximations.

Siltation of coastal marine areas can have serious consequences. Studies of the effects of higher sediment loads on coral indicated that a threshold point would be reached in some areas. Exposure of coral to slightly higher concentrations of sediment would be lethal. Below the threshold, the coral's self-cleaning mechanisms are effective. Damage to coral reefs reduces the stocks of fish which depend upon the reef. Reef fish are not directly killed by high turbidity, as they are able to escape to other areas of the reef. Mortality is indirect, resulting from a loss of habitat and food.

A dose-response model was constructed to assess the effects of sedimentation on live coral cover and fish biomass. According to regression equations derived by Hodgson (1988):

- for every 400 tons/sq km of annual sediment deposition in Bacuit Bay there is a 1% reduction in coral cover;
- for each increase of 100 tons/sq km annual sediment depositon there is annual extinction of one coral species in Bacuit Bay;
- for each 1% annual decrease in coral cover, fish biomass is decreased by 2.4%;
- for each annual decrease of one coral species associated with coral cover loss, fish biomass decreases by 0.8%.

These relationships permitted estimation of the long-term effects of sedimentation on fish diversity, biomass and catch, as shown in Figure 8.4.

Under Option 2 it was assumed that logging of the remaining area of the drainage basin would be completed within 5 years. The forest would therefore be cut at the rate of 7.4 sq km/yr, as opposed to 4.8 sq km/yr in the study period, causing sediment deposition in Bacuit Bay to increase by a factor of 1.54. This assumes that sedimentation is proportional to the rate of logging. The rate of decrease of erosion after logging has ceased was estimated to be 30% per annum, starting from year 2. Discharge of sediment from the Manlag River and deposition in Bacuit Bay was assumed to remain in proportion to the logged area until the second year. Lastly, it was assumed that the entire bay and shelf fisheries catch would decline at a rate established by the dose-response equations for coral and fish, given a constant rate of fishing effort.

Under Option 1, the banning of logging operations in the drainage basin of Bacuit Bay would prevent future damage to the catchment area and arrest the damage to coral reef ecosystems already taking place.

Effects on Financial Revenues of Economic Activities in Bacuit Bay

Gross revenues of the three industries were estimated for Options 1 and 2 over a ten year period. It was assumed that this period of analysis would limit magnification of potential errors in predictions and underlying assumptions. The present values of gross revenues for each of the three industries were then calculated and compared. The analysis was carried out in constant 1986 prices. Discount rates of 10% and 15% were applied.

Tourism in Bacuit Bay consists of two scuba diving resorts and various other smaller scale businesses. Estimates of gross revenue from tourism were calculated using information on mean length of visitor stay, mean occupancy rate and advertised daily rates for each resort and for another category of tourist activity.

Under Option 1 the tourism sector is expected to expand in the years 1991 through 1996. Under Option 2, tourism revenues are reduced by 10% per annum in the years 1988-1991 as a result of the degradation of dive sites. From 1992 through to 1996, gross tourism revenues for the area are reduced to the 1992 value of the "other" category because the two scuba diving resorts are assumed to go out of business as a result of environmental deterioration.

To assess the effects of Options 1 and 2 on fishing in Bacuit Bay, several assumptions were made. The predicted fish catch and revenue under Option 1 was based on actual 1986 catch and gross revenues. As the 1986 fish catch, according to local and commercial fishermen, was considered below normal, the estimated fish catch in later years under Option 1 was assumed not to decline further. Delayed impacts of logging carried out before 1985 were assumed to cancel out over the ten year period of analysis. The catch of large commercial fishing boats was adjusted downwards to reflect the fact that these boats spend only a proportion of their fishing effort in the vicinity of Bacuit Bay.

Under Option 2 the predicted reduction in fish catch was calculated using the dose-response regression equation for coral cover, species diversity and fish biomass (see Hodgson and Dixon 1988).

For continued logging operations, as under Option 1, the gross revenue falls to zero after logging is completed. For Option 2, annual gross revenue estimates were calculated using price data from international sources and estimates of production and mill wastage.

A summary of the present values of gross revenues for all three industries in the Bacuit Bay area, using discount rates of 10% and 15%, is given in Table 8.2. The results indicate that, under the assumptions adopted, Option 1 is economically superior at both discount rates.

Sensitivity Analysis and Interpretation of Economic Analysis

Sensitivity analysis was used to test the results of the economic analysis in response to changes in values of key parameters. Three alternative scenarios were examined. Scenario A tests the sensitivity of the results to changes in the predicted losses to tourism and fisheries as a result of logging. The predicted annual losses of revenue from fisheries and tourism were assumed to be half the

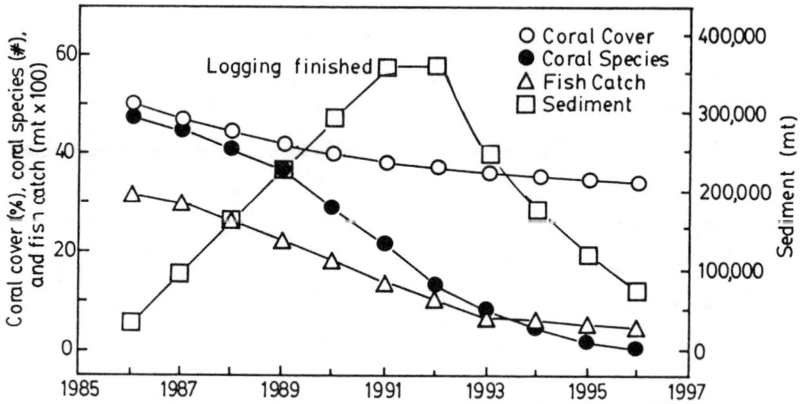

Figure 8.4. Effects of logging on coral and fish catch in Bacuit Bay.
Source: Hodgson and Dixon (1988).

levels used in the original analysis. The results of this sensitivity analysis did not alter the conclusions of the study.

Scenario B tests the sensitivity of the original results to assumptions about predicted expansion in the tourist industry between 1991 and 1996. Under Scenario B there is no expansion in the tourism industry despite the logging ban under Option 1. Again, the main conclusions are maintained.

Scenario C tests the sensitivity of the results to a combination of Scenario A and Scenario B. Under Scenario C the effects of sedimentation damage are reduced by half and there is no tourism expansion in the years 1991-1996. Under this scenario the gross revenues of Option 1 are higher than the gross revenues of Option 2. The present values of both options at a 10% discount rate are roughly equal. At a 15% discount rate Option 2, continued logging, has a slightly higher present value.

The results suggest that a high economic cost is associated with continued logging. The preferred economic option is to ban further logging, assuming that the tourism and fishery industries operate in an environmentally sound and sustainable manner. These results must be treated with caution since it was gross returns, not net returns, that were considered in the analysis. No account was taken of the fixed and variable costs associated with each industry. If tourism and fishing had significantly higher costs than logging, the conclusions of the analysis could be reversed.

Table 8.2.

Ten Year Sum of Gross Revenues and Present Values of Gross Revenues for Each Industry Sector Under Option 1 (Logging Ban) and Option 2 (Continued Logging). Discount Rates 10% and 15%. Values in $'000.

	Option 1	Option 2	Option 1 - 2[a]
Gross Revenue			
Tourism	47,415	8,178	39,237
Fisheries	28,070	12,844	15,226
(with tuna)[b]	(46,070)	(21,471)	(24,599)
Logging	0	12,884	-12,884
Total	75,485	33,906	41,579
Present Value (10%)			
Tourism	25,481	6,216	19,265
Fisheries	17,248	9,108	8,140
(with tuna)[b]	(28,308)	(15,125)	(13,183)
Logging	0	9,769	-9,769
Total	42,729	25,093	17,636
Present Value (15%)			
Tourism	19,511	5,591	13,920
Fisheries	14,088	7,875	6,193
(with tuna)[b]	(23,122)	(13,083)	(10,039)
Logging	0	8,639	-8,639
Total	33,599	22,125	11,474

[a]This column lists the difference obtained when Option 2 values are subtracted from Option 1 values.

[b]The values for the Fisheries Sector when tuna catch is included are shown in parentheses but they are not used to calculate the totals.

Source: Hodgson and Dixon (1988)

Qualitative Considerations: Economic, Social and Ecological

The economic analysis considered only the gross revenues of the three major industries. Other benefits might accrue to the islanders if a logging ban were instituted. They include the following:

Maintenance of habitat and constituent species. Biological resources are valuable in terms of tourism and research. A ban on logging would help prevent a reduction in numbers and diversity of flora and fauna. Community benefits would flow from such effects.

Provision of infrastructure. Infrastructure provided by the tourism industry is well located to serve the inhabitants. It is also more permanent and less environmentally damaging than infrastructure designed to service the logging industry. The social benefits of infrastructure provided by tourism would therefore be greater and longer lasting.

Flood damage. Maintaining an unlogged, natural catchment helps minimise flash flooding, and consequently damage to buildings, agricultural land and infrastructure.

Employment. Studies have shown that little or no female employment is provided by the logging industry. Although male workers acquire a skill, the period of employment may be only a few years. By contrast, employment in the other Bay industries makes use of female labour and the jobs are of longer duration.

Intergenerational equity. If operated on a sustainable basis, tourism and fisheries industries would provide income for generations to come. Current logging practices in the concession area are damaging the timber resource, thus denying future generations the opportunity to derive an income even from forestry operations.

References

Asian Development Bank (undated) Appendix II Case Study Summary for Palawan.

Brillantes, A. (1992) *Environmental Regulations in the Third World (an Assessment of the EIA System in the Philippines)* 3rd Asia-Pacific Mining Conference, Manila, 19 March, 1992.

del Rosario, E.G. (1988) 'Environmental Impact Assessment in the Philippines', ESCAP Papers.

Finney, C.E. and S. Western (1986) 'An Economic Analysis of Environmental Protection and Management: An Example from the Philippines', *The Environmentalist*, 6,1: 45-61.

Hodgson, G. and J.A. Dixon (1988) 'Logging Versus Fisheries and Tourism: An Environmental and Economic Analysis', Occasional Paper No 7, Environment and Policy Institute, East-West Center, Honolulu.

Hunting Technical Services Limited, Sir M. MacDonald and Partners and Planning, Management and Development Systems, Inc. (Philippines) (1985) 'Palawan, A Strategic Environmental Plan', PIADP.

National Environmental Protection Council 1983, Environmental Impact Assessment Handbook.

Presidential Decree No 1151, Promulgating the Philippine Environmental Policy.

Presidential Decree No 1586 Establishing an Environmental Impact Statement System including Other Environmental Management Related Measures and for Other Purposes (1978).

Quisumbing Torres Evangelista (date unknown) *An Introduction to the Philippine Environmental Impact System*, Manila.

Rules and Regulations Implementing Presidential Decree 1586 in relation to Presidential Decree 1151 (1979)

'Strategic Environmental Plan for Palawan, Integrated Environment Program', PIADP, October 1985.

Chapter 9

FLOOD CONTROL IN THE KAHAWAINUI WATERSHED, ISLAND OF OAHU, HAWAII

DAVID JAMES

Hawaii is the fiftieth State of the United States of America. It came into the Union in 1959, having been a United States Territory from 1900. Its population is approximately 1.6 million, with a significant indigenous component. The distribution of land use is 5% urban, 47% agriculture, 47% conservation and 1% rural.

Kahawainui Stream is located on the northeast coast of the Island of Oahu in the State of Hawaii (see Figure 9.1). The watershed contains the town of Laie, the Brigham Young University-Hawaii Campus and the Polynesian Cultural Center. The area has significant scenic, cultural, ecological and recreational values and constitutes a major tourist attraction.

The watershed is subject to frequent flooding, causing widespread damage. As of 1985, no flood protection works had been undertaken in the area, despite several earlier studies of the flooding problem.

The EIS reported in this case study was prepared in 1985 by the US Army Corps of Engineers, Pacific Ocean Division, Honolulu District. The EIS investigated the areas subject to flooding, causes of flooding, possible measures for flood protection, and benefits, costs and environmental impacts associated with protection measures. The purpose of the study was to identify the problems and needs of flooding in the area, and assess the extent to which the Federal

D. James, The Application of Economic Techniques in Environmental Impact Assessment, 247–274.

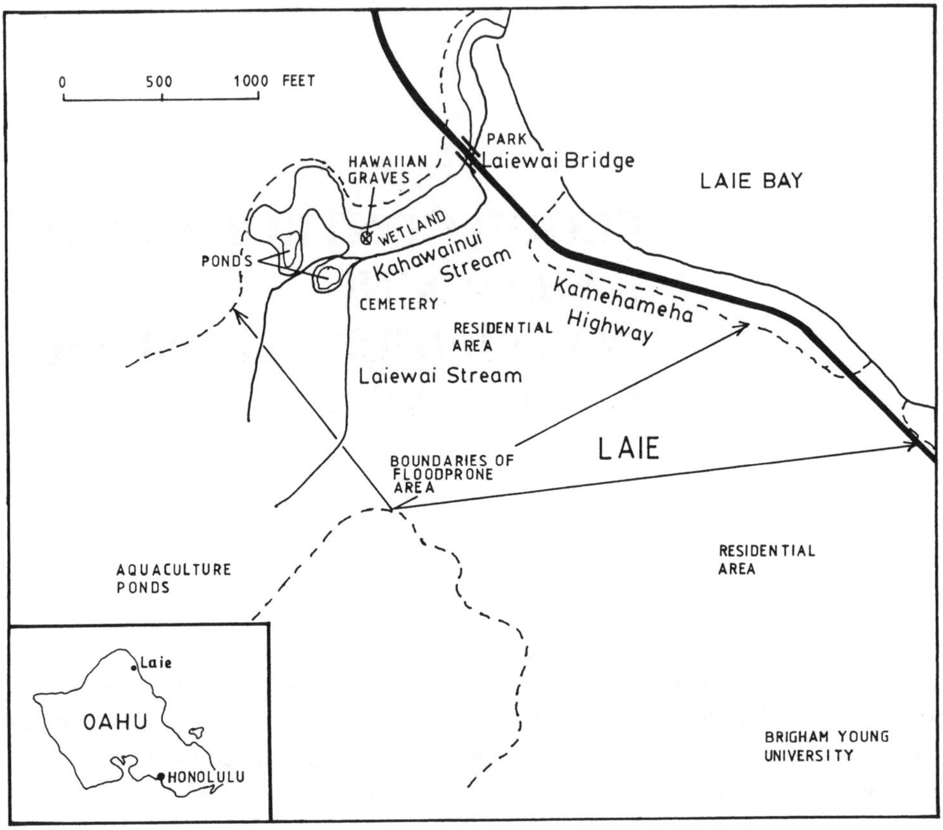

Figure 9.1. Map of Kahawainui Stream study area

Government should participate in flood damage reduction measures. The planning period extends from 1990 to 2040.

The EIS is an excellent example of the application of economic risk analysis to flood control, as routinely practised by the US Army Corps of Engineers. It also shows how economic information can be combined with environmental accounts to facilitate project evaluation and planning.

Description of Study Area

General

In keeping with the procedures prescribed for the preparation of Environmental Impact Statements by the Army Corps of Engineers, an investigation was made of significant resources in the study area. These were described briefly in the final EIS, with supporting details in technical appendices. The main features are reproduced below.

Physical Environment

The Kahawainui watershed consists of five sub-drainage areas covering an area of 5.3 sq miles. The boundaries of the watershed are defined by the Koolau Mountains rising to a height of 2,000 feet on the western side, and beach berms along the eastern flank. Drainage outlets to the sea are limited. Interior ponding occurs in low lying areas, due to the physiographic features of the floodplain. The watershed contains two drainage ponds as well as seven ponds used for aquaculture. Median annual precipitation varies from 45 inches on the coast to over 100 inches in the mountain areas.

The watershed contains Prime Agricultural Land and Other Important Agricultural Land as designated by the State of Hawaii. Forest reserve land accounts for about 60% of the total area, with pastoral land, mixed agriculture and urban development comprising the rest.

Plant species are predominantly exotic. There are some remnants of indigenous vegetation, but none of the native plant species can be classified as threatened or endangered.

Wetlands occur within the study area. There are two ponds with surrounding marshland, a fringing marsh and aquatic vegetation adjacent to Kahawainui Stream itself, and mangroves and hau swamp in the lower estuarine parts of the catchment. The wetlands filter sediments and nutrients, provide a buffer zone against storm waves and flood flows, recharge aquifers and provide recreational fishing and scenic amenities. The wetlands also constitute a habitat for waterbirds, including the Hawaiian Gullinule, an endangered bird species.

Terrestrial fauna in the study area are mostly exotic species, especially dogs, cats, rats and mongooses.

Aquatic species in Kahawainui Stream are not diverse. In field surveys no invertebrates could be found. Fish species are restricted to the estuarine section and include mullet, poeciliids and tilapia.

Poor water quality is the main reason for the limited diversity of aquatic life. The stream and adjacent wetlands are naturally characterised by low levels of dissolved oxygen, and hydrogen sulphide is produced by black, silty bottom sediments. There are also high concentrations of ammonia from the use of agricultural chemicals. Kahawainui Stream is designated a Class 2 stream in accordance with the classification scheme used by the State of Hawaii. This designation should protect water quality for recreation, propagation of fish and other aquatic organisms, and use of water for agricultural and industrial purposes.

Sediment discharges from Kahawainui Stream to coastal waters do not significantly affect marine life. Coastal currents and wave action adequately dissipate sediment loads. The adjacent coastal waters are designated Class A by the State of Hawaii.

Recreational and Aesthetic Values

Recreational activities in the area are provided by Hukilau Park, a 9-acre reserve located between Kamehameha Highway and the coast, near the mouth of Kahawainui Stream. The park is a low lying area used for picnicking, camping, and recreational fishing. Extensive areas of ironwood trees and coastal mangroves, coastal vistas and offshore islands contribute to the scenic quality of the area.

Social and Economic Environment

The socio-economic history of the region was traced to the 1850s when Morman missionaries first settled in the area. By 1865 the Morman Church had acquired 6,000 acres of land and commenced agricultural activities. Provision of pumped water boosted agriculture towards the end of the century, with sugar cane and pineapples comprising the major crops. Sugar production remained the predominant form of agriculture until 1970 when the last plantation closed. Agriculture now consists of mixed farming, grazing and aquaculture.

The resident population of Kahawainui watershed has grown from 3,810 in 1970 to 5,752 in 1980. Most of the population is located in Laie. The average Laie household consists of 5.8 persons. Incomes tend to be low, with 43% of households earning less than $10,000 per year.

Economically, the area depends heavily on the tourist trade. Since 1959 tourist expenditure in the State of Hawaii has grown at an average annual rate of 17%, and in 1980 it exceeded $2.8 billion. Local employment is currently supported by the Brigham Young University and the Polynesian Cultural Center, both controlled by the Morman Church. General services are available in the Laie urban area. Kamehameha Highway, which follows the coastline of Oahu, provides an important link between Laie and the City of Honolulu.

Cultural Resources and Historic Sites

Various cultural resources discovered in the Kahawainui watershed could meet the requirements for inclusion in the National Register of Historic Sites. One is the site of an inland fishpond known as Pa'eo fishpond which dates back to prehistoric times. The pond has since been covered by alluvial sediments and no surface traits remain.

Laie Cemetery covers an area of 3.5 acres. The nearby area contains historically significant remnants. A red torii gate is the only reminder of a Shinto shrine. North of the gate are old sand dunes which may contain ancient artifacts, but these lie outside the study area.

Within the study area, north of Laie Cemetery, are the sites of old Chinese and Hawaiian graves. Close to the graves is a limestone escarpment in which numerous historic, possibly prehistoric, remains have been found.

Sand dunes on the seaward side of Kamehameha Highway, marking the site of an ancient shrine known as Hanapepe Shrine, were considered a likely source of ancient artifacts, although surface searches and stratigraphic tests failed to produce any evidence.

The Flood Problem

Frequency of Floods

Flooding occurs frequently in the Kahawainui watershed. It is caused by overbanking of Kahawainui Stream, aggravated by sheetflows which are produced when precipitation exceeds the drainage capacity of the Laie urban area, and by the accumulation of sediment and debris in drainage channels. Flood flows are usually of short duration, with rapid buildup and recession due to the nature of storms in the area and the topography of the drainage basin. Flood heights up to six feet have been experienced in recent years. Discharges near the mouth of the stream have been estimated to vary from 1,600 cubic feet per second (cfs) with a 2-year flood, to 17,800 cfs with a 285-year flood.

Since 1879, twelve major floods have occurred causing extensive damage. Five major flood events occurred between 1963 and 1972. Flood damage has included the loss of homes and crops, emergency evacuations, washing out of roads and bridges, and disruption of telephone and power facilities. In the widespread flooding that took place on Kauai and Oahu in 1963, the Kahawainui floodplain was one of several areas declared as disaster areas. Federal aid of $250,000 was granted to help repair the damage. In 1982, flood damage exceeding a value of $630,000 was caused.

Legal and Institutional Setting

(Ben Boer and Angus Martyn)

Environmental Impact Assessment Requirements

Hawaii, in common with other States of the United States, must conform to all federal requirements as well as to State requirements in the area of environmental assessment. The *National Environmental Policy Act* of 1969 applies to all federal agencies which have jurisdiction over development proposals in any part of the United States. The Act requires all federal agencies to prepare environmental impact statements for all major actions over which they have jurisdiction that may significantly affect the environment. The statement, according to Section 102(c) must include:

- the environmental impact of the proposed action;
- any adverse environmental effects which cannot be avoided should the proposal be implemented;
- alternatives to the proposed action;
- the relationship between local short-term uses of the human environment and the maintenance of productivity; and
- any irreversible and irretrievable commitments of resources which would be involved in the proposed action should it be implemented.

The Council for Environmental Quality, also set up under NEPA, and directly responsible to the President of the United States, issued detailed regulations in 1978 for the carrying out of environmental impact assessment under NEPA. The regulations were designed to reduce paperwork and delays, and generally to encourage better EIAs.

Since the introduction of NEPA, many States have enacted their own EIA legislation, colloquially known as "little NEPAs". These statutes are designed to require EIAs in circumstances where NEPA might not apply. The *Hawaiian*

Environmental Impact Statement (HEIS) law dates from 1974. The legislation requires an EIS in relation to a range of actions. Without an EIS the project cannot proceed.

The HEIS is administered through the Office of Environmental Quality Control and the Environmental Council. The composition of the Council includes representatives from employee groups, management, industry, environmental conservation groups and the architectural, engineering and planning professions. The Council acts as an appeal body where the decision making agency produces a finding adverse to the project proponent. A public hearing can then take place on the island where the action is proposed.

Where both NEPA and the Hawaiian law require an EIS, the Office of Environmental Quality Control makes the decision on which agency prepares the EIS. It is also possible for joint EISs to be authorised (Callies 1984).

Application of EIA Legislation to Case Study

The US Army Corps of Engineers carried out the EIS on Kahawainui Watershed under the auspices of NEPA and the federal *Flood Control Act* 1948. The statute authorises the Chief of Engineers to perform investigative studies and to construct "small scale" flood control works, subject to an upper limit of $4 million. The provisions of HEIS presumably applied to the project, even though this is not explicitly stated in the EIS. The EIA process was placed under the control of federal authorities, although the local sponsoring agency was the City and County of Honolulu Department of Public Works.

Other relevant legislation includes the *River and Harbour and Flood Control Act* 1970 and the *Water Resources Development Act* 1974. At least 17 possible adverse economic, social and environmental impacts must be considered in project evaluations. For the Kahawainui study, the most relevant requirements related to Wetlands, Endangered Species, Historic Sites, Desirable Community Growth and Cohesion, and Agricultural Lands. The Corps of Engineers Policy Guidelines applied to the study, as well as the Principles and Guidelines for Planning Water and Related Resources. There were also requirements for certain types of information to be displayed in the EIS.

Methodology of Flood Control Planning

Identification of Problem and Objectives

The approach followed in the EIS is described in the Principles and Guidelines for Planning Water and Related Land Resources Implementation Studies laid down by the US Water Resources Council in 1983.

The first step involves an investigation of the base conditions in the without-project situation. Significant resources of the study area should be identified.

A projection is then made of the "most probable future" in the without-project situation, entailing projections of demographic, economic, social and environmental factors. This projection provides a basis for an analysis of planning needs which should reflect national planning policies and local public concerns.

The Principles and Guidelines of the Water Resources Council emphasise promotion of National Economic Development (NED) as the primary planning objective. NED benefits should reflect improvements in economic efficiency at the national level, measured by the increase in the value of goods and services produced. Consideration should also be given to Environmental Quality (EQ), Regional Economic Development (RED) and Other Social Effects (OSE). A balance sheet of accounts is set up to represent these additional considerations, in conjunction with the economic benefit-cost accounts.

The environmental quality analysis deals with effects of the proposed development on ecological, cultural and aesthetic attributes of the environment. The regional analysis covers effects on regional income, employment, population, economic base and social development. Other social effects are security of life, health and safety, urban and community impacts, emergency preparedness, displacement, long term productivity and energy.

Economic risk analysis of flood control options was only one required component of the full environmental impact assessment. Nevertheless, as shown for this particular study, it can become the dominant guide to plan selection.

Approach to Analysis

The benefits of flood control consist of the reduction in flood damage costs achieved by a particular structural design or set of mitigation measures, compared with damage costs in the without-project situation.

The economic damage costs corresponding to each flood event are estimated by means of an inundation-damage or stage-damage function. This yields an

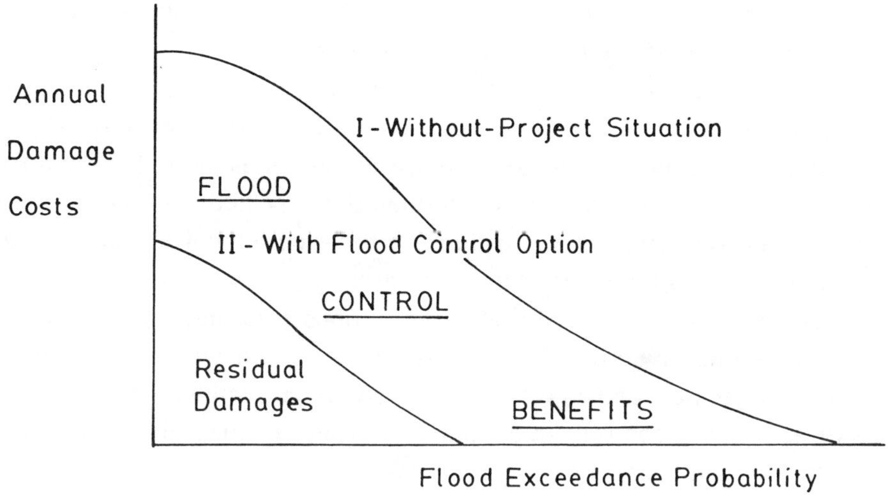

Figure 9.2. Estimation of flood control benefits.

economic valuation of the damage caused to property, crops, infrastructure etc, by each level of flooding. Event-related damage costs are translated into annual estimates by calculating annual losses such as crops, and by amortizing capital replacemement costs over the estimated life of replaced assets, using an appropriate discount rate.

An inundation-damage exceedance-probability curve can be derived by plotting the estimated annual damage costs against flood exceedance probabilities, as shown in Figure 9.2. The exceedance probability of floods is the reciprocal of the flood return period. A 100-year flood, for example, has an exceedance probability of 1/100. The inundation-damage exceedance-probability curve typically slopes downward from left to right. Its ogive shape derives from the underlying probability distribution for rainfall. Some studies use an exponential decay or similar function to represent the inundation-damage exceedance-probability curve. Floods that occur with low probability (that is, with long return periods) are characterised by large flows and high event-related levels of damage. Small floods (with short return periods) are exceeded often, but cause only low levels of damage.

Each small change in exceedance probability measures the annual probability of occurrence of the associated annual damage cost. The expected (annual average) economic value of flood damage over the full range of possible flood events is estimated as the area under the inundation-damage exceedance-probability curve. It consists of the sum of all annual damage cost levels multiplied by the corresponding annual probabilities of occurrence.

In Figure 9.2 the inundation-damage exceedance-probability curve for the without-project situation is represented by curve I. The curve corresponding to a particular engineering option for flood control is shown as curve II. The difference in area between the two curves measures the expected economic benefits of the control option - that is, the difference between annual average damage costs in the with- and without-project situations.

The expected benefits of each flood control option depends on the effect of the option on annual average damage costs. Each option will also involve an annualised construction and operating cost. By calculating the net economic benefits of each flood control option and comparing the outcomes, the best alternative (the one that maximises net benefits) can be identified.

Flood damage can still be caused, however, by floods exceeding the design level. Such damage, referred to as "residual damage", can also be calculated as an average annual value, as shown in Figure 9.2. If residual damage involves catastrophic effects, loss of life or disruption of vital services, a higher level of flood protection should be achieved, regardless of economic efficiency considerations.

The required level of flood protection depends on the severity of expected residual damages. In high density urban areas and for channels with high velocity flows, the desired level of protection is designated the Standard Project Flood (SPF). The SPF is the flood that would occur under the most severe combination of meteorologic and hydrologic conditions expected in the region. It would be exceeded only on rare occasions. The SPF can be identified using Standard Project Flood Determinations. It requires analysis of precipitation in the region, calculation of effective rainfall, estimation of surface runoff, and derivation of flood hydrographs based on sub-area hydrographs and channel routing.

Flood Control Alternatives for the Kahawainui Watershed

Without-Project Conditions

In the without-project situation, continued flooding would occur in the Kahawainui watershed. Damage could be expected to property, contents of structures, public facilities and services. The demand for housing in the area will intensify in future years as a result of population growth, although the number of structures at risk from flooding would not increase, due to new zoning restrictions introduced by the City and County of Honolulu in conjunction with the Federal Flood Insurance Program.

Plans Considered and Rejected

Several alternatives were considered but not investigated further because of adverse attributes or unacceptable economic factors. The construction of a large reservoir or a series of small detention ponds upstream was considered as a flood control measure. The capital costs of this proposal, however, would have been prohibitive.

Another rejected alternative consisted of diversion channels designed to carry flows from Kahawainui Stream to adjacent watersheds. This proposal also involved high costs, and would have caused major disruptions in wooded areas, agricultural areas and developed and undeveloped lands. Sediment loads in coastal receiving waters would have increased, resulting in a deterioration of water quality.

Description of Alternative Plans

Four Alternative Plans were investigated in detail as a means of controlling flood damage. Three of the Plans involved engineering structures. The remaining Plan entailed non-structural flood management strategies. Provision was made within each Plan for different design levels of protection to permit selection of the optimal design. The four Alternative Plans consisted of:

- Alternative 1: Levee Plan
- Alternative 2: Levee and Channel Plan
- Alternative 3: Floodproofing and Floodplain Management Plan
- Alternative 4: Channel Plan

Alternative Plan 1.

Alternative Plan 1 involved the construction of a levee along the right stream bank of Kahawainui Stream, beginning at Laeiwai Bridge on Kamehameha Highway and extending upstream for 2,370 feet. The levee would be constructed from earth fill, with concrete riprap on the river side to prevent bank erosion. To accommodate larger flows at Kamehameha Highway, concrete riprap protection of the highway embankment and additional bridge reinforcements would be provided. A concrete wall would be built to avoid relocation of

Laie Cemetery. Construction of the levee would result in flooding of areas on the left side of the river previously flood-free in the without-project situation. Additional flood protection would be required for three residential properties subject to this new flood risk. The park downstream from Kamehameha Highway would continue to be flooded.

Alternative Plan 2.

Alternative Plan 2 was a combination of channel modification and the construction of levees. The existing stream would be replaced by a channel upstream from the highway, lined with concrete. Levees of compacted earth fill would be built along the right side of the bank, with a length of 1760 feet and a maximum height of 12.6 feet. Bridge modifications would also be necessary. One residential structure and three sheds would have to be relocated. Flood protection would be provided for another 140 feet below Kamehameha Highway. Flows exceeding the design level of 400 cfs below this point would result in sheet flows to the ocean. The park below the highway would remain a flood hazard.

Alternative Plan 3.

Alternative Plan 3 was a non-structural plan involving floodproofing, zoning and floodplain management. Approximately 400 structures would be affected. Depending on the level of protection provided, 139 to 162 structures would be raised above flood level, 60 to 96 would be provided with closures or cover panels to waterproof them during times of flood, and 118 to 179 would be replaced with new structures with first floors above the flood level or be subject to major structural alterations. Between 10 and 28 structures would be protected by a flood wall. These measures would be supplemented by a floodplain management plan prohibiting new building in areas at risk, special maintenance programmes, and flood warning systems.

Alternative Plan 4.

Alternative Plan 4 consisted of a straight channel 3,850 feet long extending from a point 136 feet below Laiewai Bridge to upper Kahawainui Stream. The channel would be trapezoidal in shape and have a concrete riprap lining. The stream would be deepened at Laiewai Bridge and lined with concrete riprap. Bridge modifications would also be carried out. Parts of the park downstream from the highway would be subject to flooding.

Impacts of Alternative Plans

Ecological Impacts

Effects on Wetlands. Under Alternative Plan 1, floodwaters would be diverted to wetland areas. This would not significantly affect existing vegetation, as it is well adapted to flood conditions. Alternative Plan 2 would result in the disposal of landfill to wetland areas, destroying about 2 acres or 10% of the wetland area in the watershed. Alternative Plan 3 would have no effect on wetland areas. Alternative Plan 4 would have significant effects on the wetland environment, but as a result of channelisation, flows through wetland areas would be accelerated, disrupting normal ponding processes and possibly lowering the normal water levels.

Endangered Species. It was not possible to assess whether the proposed Alternative Plans would adversely affect the use of wetland areas by the Hawaiian Gullinule. Construction activities would have only a minor temporary impact, but the long-term effect of changes in waterflows on the Gullinule could not be predicted. The US Fish and Wildlife Service concluded that the project would not threaten existence of the species, but nevertheless issued strict guidelines on construction of engineering works that would minimise risks for the Hawaiian Gullinule. Other waterbirds visiting the area would not be threatened by the proposed developments.

Water Quality. Under Alternative Plan 1, levee construction could lead to increased sedimentation in wetland areas because of flow diversions and longer settling times. Downstream water quality would improve, but periodic maintenance could be required to control sediment loads in interior ponds and wetland areas. Under Alternative Plans 2 and 4, channelisation would lead to higher flow velocities, bank erosion and increased sedimentation loads in the seaward section of Kahawainui Stream. Alternative Plan 3 would leave water quality unaffected.

Aquatic and Marine Organisms. Sedimentation accompanying the channelisation in Alternative Plans 2 and 4 was considered unlikely to affect adversely the marine environment, due to dilution and clearing currents. Channelisation could, however, lower the water level in wetland areas and adversely affect aquatic organisms. Alternative Plans 1 and 3 would have a neutral or beneficial effect on aquatic and marine organisms.

Impacts on Historic Sites

Alternative Plans 1 and 3 were predicted to leave historic sites unaffected. Alternative Plan 4 would, without modification, encroach on the Pa'eo fishpond site, sand dunes possibly containing cultural artifacts, the Shinto torii gate and nearby graves. Potential impacts of Alternative Plan 2 would be avoided through careful design. Special measures were prescribed to eliminate the need to relocate Laie Cemetery. An archeological monitoring programme was recommended to prevent loss of cultural artifacts that might be unearthed by construction works in the proximity of the sand dunes close to the coast, Laie Cemetery and the Hawaiian and Chinese grave sites.

Effects on Community Growth and Cohesion

Alternative Plan 1 would encourage community growth and cohesion by reducing flood damage in the Laie urban area. Public health hazards would also be reduced. A higher risk would be incurred of a washout of Laiewai Bridge because of higher stream flows, which would disrupt community services such as access to hospital facilities. Fire prevention services would not be seriously affected. Bridge use could be restored within two or three days following a washout. Similar beneficial and adverse community impacts could be expected for Alternative Plan 2. Community effects of Alternative Plan 3 were more difficult to assess, because they would depend on specific planning measures. Social disruption could still be caused by flooding. Rezoning could also be required to manage effectively the floodplain, involving a shift in land use to cope with intermittent flood conditions. This Plan was considered to be less supportive of community growth and cohesion than the other Plans.

Effects on Agricultural Lands

Within the Kahawainui watershed, thirteen acres of Prime Land north of Kahawainui Stream are used for livestock raising, and one acre of Other Important Land south of the stream is used for mixed fruit and vegetable production. Elsewhere in the watershed, Other Important Land supports grazing animals. Alternative Plan 1 would divert flood waters to agricultural lands north of the stream. This would disrupt crop production, but have only a marginal effect on grazing. Alternative Plan 2 would cause a small loss of land used for banana growing near Kamehameha Highway. Alternative Plan 3 would not affect agricultural resources. Alternative Plan 4 would result in the loss of about 4.5 acres of Prime

Land used for grazing, owned by a native Hawaiian family, and have adverse effects on the use of other agricultural land in the watershed.

Summary of Impacts

A summary of the impacts predicted for all Alternative Plans is given in Table 9.1. These impacts had to be evaluated in conjunction with National Economic Development benefits in assessing the flood control options associated with each of the Alternative Plans.

Table 9.1.
Impacts Generated by Alternative Plans.

Impacts	Alternative Plan			
	1	2	3	4
Loss of Wetland Areas		X		X
Relocation of Graves				X
Impacts on Known Archaeological Resources				X
Relocation/Modification of Residential Structures	X	X	X	X
Requirement for Lands, Easements and Rights-of-Way	X	X		X
Increase Flooding to Other Areas	X			
Overtopping of Laiewai Bridge and Kamehameha Highway			X	

Detailed Economic Assessment of Flood Control Options

General Assumptions underlying the Analysis

All economic results of the investigation are expressed in 1985 prices. Some of the analyses were conducted in 1981 values, and price index corrections were required to standardise the results. To obtain annualised values for damage to property, a 50-year planning horizon was adopted and a discount rate of 8.375%

applied. The same assumptions were used to derive annualised values of flood control capital, maintenance and operating costs.

Inundation Damage Analysis

Most of the damage occurring in the Kahawainui watershed consists of damage to residential property. Of the 441 units in the floodplain, 434 are residential. The only non-residential units consist of a school, a church, a restaurant, a gas station and some food stores.

A damageable property inventory was compiled to indicate the value of buildings and contents at first-floor elevation. Structures were valued from a field survey covering 87% of all damageable units together with information from the State of Hawaii Department of Taxation on assessed values of properties. Market values of properties were estimated from assessed values, with price adjustments to convert all values to 1985 prices.

Contents of residential structures were valued on the basis of data collected in Hilo, Hawaii, which was considered to constitute a similar residential area. Contents were valued according to the ratios of contents to the market value of structures. Allowance was made for an increasing accumulation of contents over time. Contents of non-residential properties were valued separately.

In the damageable property inventory, each individual property was characterised in terms of: type of use (residential or non-residential); stream station number (coinciding with the hydrology analysis); damage code (low or high velocity flow damage zone); first floor elevation (feet above mean sea level); building type (wood on post and beam, wood on slab, brick or block on slab); assessed value of structure; and market value of contents.

Stage damage frequency functions were next developed, making use of the property inventory, standardised depth-damage relationships, and hydrology and water surface profile data.

Standardised depth-damage relationships show the ratios of structural and contents damage to the total market values of properties at different depths of flooding. Depth-damage relationships applied to residential properties in the Kahawainui floodplain are shown in Table 9.2. These relationships were checked against recent flood events to verify the damage calculation model.

Water surface profile data were used for five different flood flows, ranging from 10-year floods (exceedance probability of 0.1000) to 285-year floods (exceedance probability of 0.0035 corresponding to the Standard Project Flood).

Table 9.2.

Standardised Depth-Damage Relationships

Feet of Flooding[a]	Damage as % of Total[b] Market Value	
	Structures	Contents
0	0	0
1	18	26
2	25	40
3	30	51
4	34	61
5	37	67
6	39	72
7	41	76
8	43	79
9	45	81
10	47	83

[a] Over first floor.

[b] For residential units only.

Different relationships apply for non-residential properties.

Damage Costs in the Without-Project Situation

In the without-project situation, flood damage costs were assumed to consist of the sum of the following components:

- damage to residential structures;
- damage to contents of residential structures;
- damage to non-residential structures;
- damage to contents of non-residential structures;
- emergency relief costs.

It was realised that flooding could also threaten public property and human safety, but these effects were excluded from the analysis because of difficulties of economic measurement.

Inundation damage probability estimates for structures and contents in the without-project situation are shown in Table 9.3. Flood flows are represented in terms of their annual exceedance probabilities, and damages are measured in \$'000 on an annualised basis. The damage cost estimates are expressed in 1981 prices.

Table 9.3.
Without-Project Estimated Inundation Damage-Probability Data[a] ($'000 April 1981 price levels)

Annual Exceedance Probability	Qo[b]	Inundation Damage: Residential Structures	Residential Contents	Residential Total	Non-Residential Structures	Non-Residential Contents	Non-Residential Total	TOTAL
0.1000	4	2,551	1,017	3,568	61	236	297	3,865
0.0400	7	3,442	1,424	4,868	81	354	435	5,301
0.0200	9.6	4,158	1,793	5,951	110	519	629	6,580
0.0100	12.3	4,683	2,078	6,761	124	623	747	7,508
.0035(SPF)	16.7	5,062	2,293	7,355	146	687	833	8,188

[a] Damages without the project as shown in this tabulation include the effect of both overbank streamflow as well as local drainage not from the stream.

[b] Overbank discharge in '000 CFS.

Expected annual flood damage costs (average annual damage costs) for structures and contents in the without-project situation were obtained by integrating the damage probability estimates in Table 9.3. The calculations were performed using a computer programme which calculates damage for each unit, for each flood flow, sums the damage probability data for the entire floodplain by structures and contents, and integrates these summed damage probability estimates to derive a single figure for the average annual damage cost for the area at risk. The calculated average annual damage cost in the without-project situation was converted to 1985 prices by applying an inflationary factor of 32%. Allowance was also made for an increasing real value of residential property contents in future years by assuming an accumulation rate equal to the projected growth rate of real income over the 50-year planning period.

Estimated emergency relief costs for different flood flows in the without-project situation are shown in Table 9.4. The average annual cost of emergency relief was calculated by applying the usual integration methods to the cost probability estimates in Table 9.4.

A summary of expected annual flood damage costs for the various cost categories in the without-project situation is shown in Table 9.5.

Table 9.4.

Estimated Emergency Relief Costs ($'000)

Exceedence Probability of Flood Event	Emergency Relief Costs[a]
0.1000	15
0.0400	21
0.0200	27
0.0100	30
.0035 (SPF)	33

[a] $75 per damage unit affected in the SPF flood event, and for lower levels of flooding in proportion to damages from stage damage relationships for the floodplain. Based on experience in other floodplains within jurisdiction of Pacific Ocean Division, Corps of Engineers.

Table 9.5.

Estimated Annual Average Damages in Without-Project Situation ($'000)

Residential Structures	480
Contents of Residential Structures	289
Non-Residential Structures	12
Contents of Non-Residential Structures	53
Emergency Relief Costs	2.25
TOTAL	836.25

Estimation of Damage Costs for Flood Control Measures

Flood controls considered for the Kahawainui watershed have been defined as Alternative Plans 1, 2, 3 and 4. Within each Alternative Plan, the design level of protection was varied to withstand floods of differing return periods. Alternatives 1, 2 and 4 were designed to give equivalent levels of protection at the same flood flows. Inundation damage probability estimates for structures and contents for Alternative Plans 1, 2 and 4, measured in 1981 prices, are shown in Table 9.6. The flood return periods range from 2 years to 285 years. Beyond the Safe Project Flood level of design, further flood damage was considered negligible.

Average annual costs were calculated using the usual integration procedures. As in the without-project situation, a price index correction was applied to convert all estimates to 1985 values. Projections were made of the contents value for residential structures. Emergency relief costs at different flood flows were also estimated for Alternatives 1, 2 and 4 and converted to an annual average.

Table 9.6.

Alternatives 1, 2, and 4 Estimated Inundation Damage-Probability Data ($000's)

Annual Exceedance Probability	Qo	Inundation Damage: Residential Structures	Residential Contents	Residential Total	Non Residential Structures	Non Residential Contents	Non Residential Total	TOTAL
2-Year Design:								
0.1000	3	1,914	763	2,677	46	177	223	2,900
0.0400	6	3,117	1,289	4,406	13	321	394	4,800
0.0200	8.6	3,855	1,662	5,517	102	481	583	6,100
0.0100	11.3	4,553	2,020	6,573	121	606	727	7,300
.0035(SPF)	15.7	4,946	2,240	7,186	143	671	814	8,000
10-Year Design:								
0.1000	0	0	0	0	0	0	0	0
0.0400	3	1,883	779	2,662	44	194	238	2,900
0.0200	5.6	2,907	1,253	4,160	77	363	440	4,600
0.0100	8.3	3,680	1,633	5,313	97	490	587	5,900
.0035(SPF)	12.7	4,760	2,156	6,916	137	646	783	7,700
25-Year Design:								
0.1000	0	0	0	0	0	0	0	0
0.0400	0	0	0	0	0	0	0	0
0.0200	2.6	1,580	681	2,261	42	197	239	2,500
0.0100	5.3	2,807	1,246	4,053	74	373	447	4,500
.0035(SPF)	9.7	4,080	1,848	5,928	118	554	672	6,600
50-Year Design:								
0.1	0	0	0	0	0	0	0	0
0.04	0	0	0	0	0	0	0	0
0.02	0	0	0	0	0	0	0	0
0.01	2.7	1,627	722	2,349	43	216	259	2,608
.0035(SPF)	7.1	3,308	1,498	4,806	95	449	544	5,350

Table 9.6 Continued

100-Year Design:								
0.1000	0	0	0	0	0	0	0	0
0.0400	0	0	0	0	0	0	0	0
0.0200	0	0	0	0	0	0	0	0
0.0100	0	0	0	0	0	0	0	0
.0035(SPF)	4.4	2,508	1,136	3,644	72	340	412	4,056

SPF Design:
Flooding with the SPF design level of protection is measurably insignificant, for purposes of project economic evaluation, resulting in a negligible average annual equivalent effect.

For Alternative 3, design levels of protection were specified for return floods of 50, 100 and 285 years. The relevant inundation damage probability estimates are shown in Table 9.7. Annual average costs were obtained by integrating the data in Table 9.7 with due allowance for price index corrections and projected changes in contents values.

A summary of estimated average annual damage costs for all Alternative Plans at different design levels of protection is given in Table 9.8.

National Economic Development benefits of each Alternative and design level of protection are shown in Tables 9.9 and 9.10 (table 713.533-1). These are expressed in average annual terms and were calculated by subtracting the damage cost estimates for each possible flood control measure from the corresponding damage cost estimate in the without-project situation. Table 713.533-1 is required by Federal regulations to be displayed.

Residual Damages

Average annual residual damages associated with each Alternative Plan are shown in Table 9.11. Estimates were derived only for the 50-, 100- and 285-year design levels of protection. In the EIS it is pointed out that even though Alternatives 1, 2 and 4 theoretically provide the same levels of flood protection, Alternative 1 (which involves the construction of levees) is in fact more risky than the others. The reason is that overtopping can lead to levee failure, creating a flood wave with disastrous consequences. To minimise the risk of flash flood phenomena, levees must be carefully designed and constructed.

Table 9.7.
Alternative 3 Estimated Inundation Damage-Probability Data ($'000)

Annual Exceedance Probability	Inundation Damage[a] Residential[b] Structures	Contents	Total	Non Residential[c] Structures	Contents	Total	TOTAL
50-Year Design:							
0.1000	255	101	356	6	24	30	386
0.0400	344	142	486	8	35	43	529
0.0200	416	179	595	11	52	63	658
0.0100	2,341	1,039	3,380	111	561	672	4,052
.0035(SPF)	2,532	1,146	3,676	131	618	749	4,425
100-Year Design:							
0.1000	255	101	356	6	24	30	386
0.0400	344	142	486	8	35	43	529
0.0200	416	179	595	11	52	63	658
0.0100	468	208	676	12	63	75	751
.0035(SPF)	2,532	1,146	3,676	131	618	749	4,425
SPF Design:							
0.1000	255	101	356	6	24	30	386
0.0400	344	142	486	8	35	43	529
0.0200	416	179	595	11	52	63	658
0.0100	468	208	676	12	63	75	751
.0035(SPF)	506	229	735	15	69	84	819

[a] These residual damages at or below the design level of protection are an estimated 10% of without-project damages, reflecting the fact that flooding still occurs, resulting in damages to property other than individually floodproofed structures (lawns, cars. etc.). April 1981 price levels.

[b] For flooding above the design level of protection, damage estimates for residential units reflect the fact that structures which will have been raised will have minimal flooding, whereas structures floodproofed with closures or ring walls will be subject to a major overtopping effect.

[c] For non-residential damage units, flooding above the design level of protection reflects the fact that major overtopping effects occur. None of these units will have been floodproofed by raising the structure.

Table 9.8.

Estimated Average Annual Damage Costs for Alternative Plans 1, 2, 3 and 4 ($'000)

Level of Protection	Residential		Non Residential		Emergency Relief
	Structures	Contents	Structures	Contents	
Alternatives 1, 2 and 4					
10 year	276	253	12	53	0.50
25 year	129	152	8	45	0.50
50 year	76	132	6	39	0.45
100 year	50	115	6	35	0.44
SPF	27	100	5	32	0.42
Alternative 3					
10 year	-	-	-	-	-
25 year	-	-	-	-	-
50 year	88	54	3	16	-
100 year	67	41	2	10	-
SPF	48	29	1	5	-

Flood Control Costs

The costs of each Alternative Plan and design level of protection were estimated under the following assumptions:

- all cost estimates are expressed in 1985 prices;
- the job would be performed by an Oahu contractor;
- the construction period is 18 months;
- a 20% contingency factor is applied;
- annualisation of costs is based on a planning period of 50 years with an interest rate of 8.375%;
- average annual costs include annualised construction costs, interest costs during construction, annual maintenance costs, and costs of repairing structural elements.

Annualised costs for Alternative Plans at the various design levels of protection are shown in Table 9.10.

Table 9.9.

(713.533-1)[a] Summary of Annualised NED Benefits (Flood Hazard Reduction) for Alternatives 1, 2, 3 and 4 ($'000)

	Alternative 1, 2 and 4						Alternative 3		
	2-Yr[c]	10-Yr	25-Yr	50-Yr	100-Yr	SPF	50-Yr	100-Yr	SPF
Residential									
Structures	26	204	351	404	430	453	392	413	432
Contents	17	31	36	137	157	174	189	235	248
Non-Residential									
Structures	0[d]	0	4	6	6	7	9	10	11
Contents	0[d]	0	8	14	18	21	37	43	48
Emergency	0[d]	2	2	2	2	2	1	1	1
Public Property[b]	-	-	-	-	-	-	0	0	0
Human Safety[b]	-	-	-	-	-	-	-	0	0
Total Project Benefits	43	242	502	583	630	672	674	715	752

[a]Table 713.533-1 is required by Federal Regulations to be displayed. See Federal Register Vol. 44, No. 242, Rules and Regulations, 14 December 1979. (Water Resources Council Procedures for Evaluation of National Economic Development Benefits and Costs in Water Resources Planning). April 1985 price levels.

[b]While these effects will most likely occur, no credit is included due to measurement problems.

[c]Alternative 1 only. This low-level-of-protection levee project is examined since it involves low-cost levee construction utilizing existing topographical features as a part of the protection.

[d]Negligible effect.

Table 9.10.

Summary of Annualised NED Benefits and Costs for Alternatives 1, 2, 3 and 4 ($'000)

Design level of protection	Alternative	B[a]	C[a]	B/C[a]	B-C[a]
2-Year	1	43	97	0.44	-54
10-Year	1	242	259	0.93	-17
	2	242	211	1.15	31
	4	242	700	0.35	-458
25-Year	1	502	279	1.80	223
	2	502	240	2.09	285
	4	502	750	0.66	-248
50-Year	1	583	300	1.94	283
	2	583	262	2.23	321
	3	674	1,464	0.46	-790
	4	583	769	0.76	-186
100-Year	1	630	318	1.98	312
	2	630	283	2.23	347
	3	715	1,890	0.38	-1,175
	4	630	796	0.79	-166
SPF[b]	1	672	342	1.96	330
	2	672	317	2.12	355
	3	752	2,059	0.37	-1,307
	4	672	831	0.81	-159

[a]B = average annual benefit, C = average annual cost, B/C = benefit-cost ratio, B-C = net benefits.

[b]For purposes of economic evaluation, providing 500-year design level of protection is not significantly measurably different from SPF level of protection.

Table 9.11.

Average Annual Residual Damages

Level of Protection	Probability of an Event Exceeding the Design	Alternative Plans[a] ($'000)			
		1[b]	2[b]	3	4
SPF (285-year)	16%	164	164	84	164
100 year	39%	206	206	121	206
50 year	64%	253	253	162	253

[a] Includes interior drainage flooding damages.

[b] Excludes levee failure, assumes controlled levee overtopping.

Interpretation of Economic Results

Selection of Optimal Design

The economic analysis indicates that Alternative Plan 2 at the SPF level of protection maximises net benefits. The expected value of annual net benefits is $355,000 and the benefit-cost ratio is 2.1. This design option was considered not to have any significant adverse or non-mitigative effects on the environment, and was thus selected as the preferred National Economic Development Plan. In the judgement of the investigators, the combined beneficial NED and environmental effects of the recommended design would outweigh those of all other options.

The cost of the recommended flood control measure was estimated as $3,290,000. In keeping with the apportionment of costs under traditional policy, the US Army Corps of Engineers would bear costs of $2,760,000 and the remaining $530,000 would be covered by the City and County of Honolulu.

Sensitivity Analysis

The sensitivity of the economic results to various parameters and components of the analysis was investigated to determine whether further detailed study would be required in post authorisation planning of the project.

Conditions that would render the project economically infeasible were investigated first. A possible cause is an overstatement of flood damage costs or, equivalently, flood protection benefits. For example, the value of structures and contents, which were calculated in relation to the market values of properties, might have been overstated in the study. Continued flooding in future years could result in a decline in property values, requiring a downward adjustment of flood protection benefits.

To make any economically feasible option infeasible, the corresponding benefit-cost ratio would have to be reduced to less than 1. Sensitivity analysis showed that, for options with a benefit-cost ratio in the range of 1.8 to 2.2, property values would have to fall on average by 6% to 9.5% compound per year to produce this effect. This implies a 99% reduction in real property values by the end of the 50-year period. For benefit-cost ratios around 1.15, the required decline in property values would be 1.53% compound per year or 95% by the end of the planning period. None of these outcomes was considered very likely, suggesting that economically feasible flood control measures identified in the study would remain feasible even with a wide variation in flood protection benefits.

Flood protection benefits, however, were found to be highly sensitive to possible errors in elevation data for structures. An error of only plus or minus one foot in elevation data for all structures in the floodplain would lead to a 39% change in estimated benefits. Thus, if surface water profiles were one foot higher or the elevation of all structures one foot lower than assumed in the study, flood protection benefits would be overstated by 39%. An error of 0.5 foot would involve a variation of 19% in protection benefits. To minimise the possibility of such errors in the analysis, consistent data collection procedures were followed for all structures in the floodplain, using the same reference system as in the hydrologic analysis.

Variations in the discount rate, within a realistic range, were found not to affect greatly the results of analysis. Benefits barely respond, as the only benefits changing over time are the real values of contents of residential properties, and these comprise only a small proportion of total project benefits. Changes in the discount rate, however, do affect control costs and the corresponding benefit-cost ratios. To render economically feasible flood control options infeasible, the discount rate would have to be increased substantially. For example, control options with benefit-cost ratios around 2.0 would require an increase in the discount rate to 17% to become uneconomic, and options with ratios of around 1.2 would need a discount rate of at least 10%.

Another factor potentially affecting the results is "freeboard" or the vertical difference between the design water surface and the top of a channel or levee. For all flood control measures considered in the investigation, a minimum freeboard of three feet was assumed. To derive an estimate of freeboard benefits, informed judgement should be used, but in the absence of better information, a rule of thumb measure is one half of the area under the damage probability curve between the design level of protection and the largest flood that could occur without overtopping the channel wall or levee.

Inclusion of freeboard benefits in the calculations would, in general, shift the optimal design to a lower level of flood protection. Furthermore, freeboard effects tend to be more significant at low levels of protection, as compared with the SPF level. The investigators did not have sufficient confidence in their estimates of freeboard benefits to take them into account in their economic calculations and thus excluded them from the analysis.

References

Callies, D.L. (1984) *Regulating Paradise: Land Use Controls in Hawaii*, University of Hawaii Press, Honolulu.

Council on Environmental Quality (1983, 1985) *Environmental Quality: Annual Reports*, Washington, DC.

Environmental Quality Commission and Environmental Impact Statements, Hawaii Revised Statutes Ch 343 (1976; Supp 1980).

Grad, F. (continuous update service) *Treatise on Environmental Law*, Matthew Bender, New York, Ch. 9.

National Environmental Policy Act 1969. (Federal).

National Environmental Policy Act Regulations 1978 (Federal).

US Army Corps of Engineers (1985) 'Kahawainui Stream, Island of Hawaii, Final Detailed Project Report and Environmental Impact Statement', Department of the Army, US Army Engineer District, Honolulu.

Chapter 10

ASSESSING CUMULATIVE IMPACTS OF ALUMINIUM SMELTING IN THE HUNTER VALLEY, NSW, AUSTRALIA

GEOFF GOLDRICK AND DAVID JAMES

The traditional environmental impact statement is a useful and widely used tool for development and environmental management. Its strength lies in its ability to identify local, direct effects of specific proposals. It can be extended to address regional and indirect effects, but as the scale and complexity of the assessment increase, so does the investigational cost.

Regional systems analysis represents a more cost-effective way of analysing the regional cumulative effects of development. Regional models can assess the assimilative capacity and economic effects of different development and environmental protection strategies. Optimal strategies can be determined in accordance with economic and environmental criteria, and specific development proposals can be evaluated in terms of these criteria. Such a process avoids duplication of effort and the need for individual proponents to collect and process large amounts of baseline data.

This case study examines an individual EIS, prepared in 1980 for the Tomago Aluminium Smelter in the Hunter Valley, New South Wales, Australia (Croft and Associates 1980). The EIS is an excellent example, among other things, of the application of regional economic impact analysis to a development proposal.

D. James, The Application of Economic Techniques in Environmental Impact Assessment, 275–298.

The case study also includes a description of a regional systems analysis model constructed as part of a research project funded by the Australian National Energy Research, Development and Demonstration Council (James et al. 1983; James and Chambers 1986). The model is based on a regional input-output model and a physical model of air pollution. The two air pollutants analysed in the model were sulphur oxides and fluorides.

The EIA and modelling work were carried out independently in the 1980s. New developments have since occurred, including construction of the Tomago smelter, new power stations and industrial plants. Despite these changes, the approaches undertaken remain instructive examples of how EIA and integrated modelling can be applied to the tasks of assessing economic impacts and cumulative environmental impacts of development on a regional scale.

Regional Setting

The Hunter Region is delineated by the catchment of the Hunter River and its tributaries. Mountains to the north, west, south, and north-east enclose a valley of approximately 31,000 sq km. The Hunter River flows to the Pacific Ocean through the port of Newcastle, in the south-east of the valley. The topography results in a close correspondence between the watershed and the airshed, though the latter includes the Lake Macquarie area which contains several electric power stations (see Figure 10.1).

The Hunter Region has a population of approximately 470,000 supported by an economy based on coal mining, agriculture, electricity generation, viticulture and manufacturing. The majority of the population lives in the Cities of Lake Macquarie and Newcastle, including the surrounding suburbs. Other major population centres include Maitland, Cessnock, Muswellbrook and Singleton.

Power stations in the region service energy intensive industries such as aluminium smelters, and heavy manufacturing industries such as chemicals, iron and steelmaking. During the 1980s, approximately 11 m t of coal were used each year for the generation of electricity.

The steady increase in mining activity and heavy industry has led to the development of an extensive infrastructure consisting of roads, railways, ports and loading facilities and water supplies. Good communications and proximity to markets have attracted numerous small industries which supply services and maintain the core industries and associated work force.

The region now has two aluminium smelters. The first is at Kurri Kurri. This smelter has a capacity of 135,000 tonnes of refined aluminium per year.

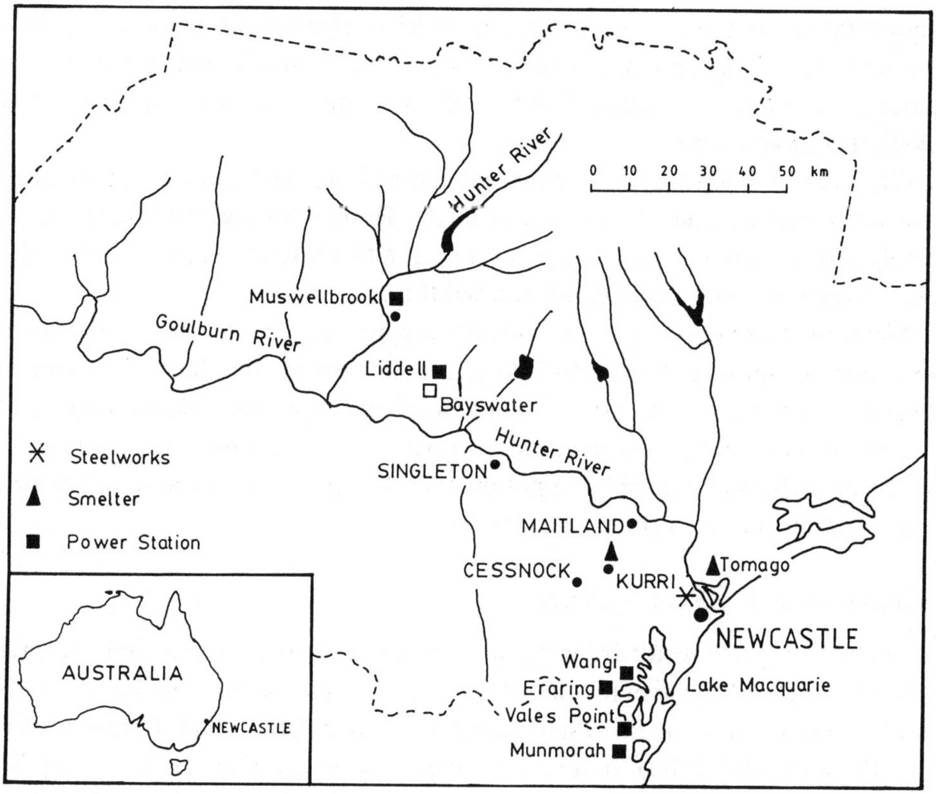

Figure 10.1. Map of Hunter Valley.

Previously its capacity was 45,000 t/yr. The second smelter is at Tomago, with a capacity of 220,000 t/yr. The case study focuses on the EIS prepared in 1980 for the Tomago venture (Croft and Associates 1980).

Environmental Issues in the Hunter

Air Pollution in the Hunter

An important environmental issue in the Hunter is the dispersion and effects of gaseous emissions from industries in the region. The narrow valley forms an enclosed airshed into which the polluting gases of a variety of industries are released. At the time of the EIA, emissions of sulphur dioxide, a by-product of coal burning and other combustion processes, amounted to 215,000 tonnes per annum. Power stations were the source of 85% of the total amount of sulphur dioxide released. At certain concentrations, this pollutant is capable of causing damage to buildings, vegetation and human and animal health. It is also a major constituent of acid rain.

Fluorides, released chiefly from power stations and aluminium smelters, have been another source of concern in the Hunter Valley. Fluorides can adversely effect vegetation and are perceived as a threat to the Hunter Valley vineyards, livestock, native vegetation and wildlife.

Many of the energy-intensive industries which are located in the Hunter region have negotiated large electricity supplies from the State Government. There have been heavy pressures on a natural environment which is required to provide not only additional quantities of coal, and hence land, but also assimilative capacity for the atmospheric pollutants released as a by-product of electricity generation and other industrial activities.

Cumulative Environmental Effects

Cumulative impacts result from the interactions of many incremental activities, each of which may have minor effects on the environment in isolation, but which become of concern when coupled with existing and future projects (CEARC and NRC 1986). In a case-by-case assessment there may be only limited knowledge of the environment's capacity to absorb impacts, and identification and control of cumulative impacts can be difficult. Often, it is only when an impact becomes damaging that action is taken to modify the proposal or restrain further development.

An effective approach is one where the environmental tolerances of the system are assessed before large-scale development occurs, so that preventive action can be taken to control enviromental impacts. Development can be kept below rates and levels considered to be environmentally damaging. In addition, the burden of pollution control can be spread more rationally among developers, to minimise the total regional costs of abatement.

The cumulative effects of incremental activities may take many forms. They may be additive or synergistic. Various types of cumulative impact may be short-term, long-term, immediate, or delayed. Cumulative impacts also have a spatial dimension.

Cumulative impacts may arise from the technical and economic linkages between a new project and other new or existing activities in a region. In the case of a new aluminium smelter, meeting the requirements for electricity may necessitate an expansion of generating output and capacity, increasing the quantities of air pollutants discharged into the region's airshed. Other environmental impacts may be caused by an increase in coal production to meet the additional demand for electricity.

In the Hunter region, cumulative environmental impacts have been a source of concern over industrial development, particularly during the phase of rapid expansion during the 1980s.

Australian Institutional and Policy Setting

(Ben Boer and Angus Martyn)

General

Australia is a federally constituted nation comprising six States and two major Territories. The population is approximately 17 million, with about 85% living in urban areas. There is a large primary industry sector with a substantial mining component, as well as significant manufacturing, tourism and service sectors.

The Australian Constitution, which dates from 1901, divides the legislative powers between the Commonwealth Parliament and the State Parliament. The High Court of Australia is the body which determines conflicts between the States and the Commonwealth in relation to the interpretation of the Constitution, and specifically the distribution of powers between the two entities.

Environmental Legislation and Administration - Federal and State

The Australian Constitution does not specifically mention environmental and natural resource matters, and it has in the past been generally assumed that legislative power in these areas lies largely with the States. However, in recent years, the High Court of Australia has recognised the power of the Commonwealth to influence environmental and resource use decisions at State level by way of powers relating to taxation, trade, external affairs and trading and financial corporations, as well as its power to legislate in respect of Aboriginal people.

In the past 20 years, on the basis of the constitutional powers mentioned above, the Federal Government has passed a wide range of legislation relating to environmental and natural resource matters. The most important of these in the present context is the *Environment Protection (Impact of Proposals) Act* 1974. This Act enables the Federal Government to require environmental impact assessment of development proposals falling within its jurisdiction. The legislation also allows the Federal government to hold public inquiries into environmental issues.

The Federal EIA process is a discretionary one, at the behest of the minister (the "action Minister") in charge of the department most closely concerned with the development activity. The Minister for the Environment does not have a direct decision-making power in relation to the application of the legislation, although influence can be exercised through the Cabinet. The federal legislation contains no specific enforcement mechanisms and there is no general provision for "standing" to challenge decisions in the court.

Detailed implementation guidelines are provided by the Administrative Procedures which have been promulgated under the Act. These procedures were extensively amended in 1987. Environmental impact assessment takes place at Commonwealth level under a two-tiered process. Where environmental impact is likely to be less significant, a Public Environmental Report is required which may or may not lead to an environmental impact statement. Alternatively, where it is obvious that environmental impact is likely to be significant, an environmental impact statement can be required at the initial stage of consideration. The requirements for supply of information by proponents and contents of a Public Environment Report and an Environmental Impact Statement are specified in some detail in the Administrative Procedures. The proponent is responsible for arranging or for carrying out the environmental impact assessment and furnishing the Public Environment Report and/or the Environmental Impact Statement. The Act and the Administrative Procedures are implemented by the Commonwealth Environmental Protection Agency (CEPA, established in 1992) which lies within the Department of Environment, Sports, and Territories (DEST).

In the State of New South Wales, the relevant environmental legislation is the *Environmental Planning and Assessment Act* 1979, a broad-ranging and sophisticated piece of legislation which integrates the processes of planning, development control and environmental impact assessment. The Act generally applies to the activities of all government departments as well as to the private sector.

Comprehensive environmental impact assessment, involving the preparation of an EIS, is required under the Act if a proposed development falls into either one of two categories. Firstly, under Part IV of the *Environmental Planning and Assessment Act*, local government authorities have the power to decide development applications relating primarily to private sector development. Schedule 3 to the Regulations under the Act specifies a list of "designated developments", for which it is compulsory to furnish an EIS together with the development application. Alternatively, local government authorities may declare any particular development as "designated". Schedule 3 includes a wide range of extractive and manufacturing industries, major agricultural activities, chemical industries and such things as marina and canal development. Secondly, environmental impact statements can be required under Part V of the Act. Part V applies to any "activity" for which no consent is required by a local authority under part IV of the Act. "Activity" is defined by the Act to include the erection of a building, the carrying out of a work, the use of land, a building or a work, and the subdivision of land. An EIS is required in relation to any activity which is "likely to significantly affect the environment".

The regulations under the Act specify detailed requirements in relation to what factors must be taken into account in deciding what the likely impact is of a proposed development on the environment; the required contents of an EIS are also specified. Apart from a description of the proposed activity, the surrounding environment and the likely interactions between the two, the regulations state that the environmental impact statement must justify the activity in terms of environmental, economic and social considerations. It must also detail measures to protect the environment in relation to the development activity and must specify any alternatives available to the carrying out of that activity.

A significant feature of the New South Wales legislation is the power of the Minister to "call-in" any development application for determination, when he or she considers it is expedient "in the public interest to do so, having regard to matters which in the opinion of the Minister are of significance for State or regional planning". This means that the Minister may take decisions out of the hands of local planning authorities and decide them centrally. In the case of "designated development", any person who objected to the development may demand a Commission of Inquiry to look into any aspect of the development if the Minister exercises the "call-in" power.

Commissions of Inquiry are constituted by independent commissioners under the *Environmental Planning and Assessment Act* who hear evidence, receive written submissions and write extensive reports which contain recommendations to the Minister in relation to development decisions.

Another feature of the New South Wales legislation is the provision for public participation in local environmental planning, development control and environmental impact assessment. In relation to environmental impact assessment, there are extensive requirements for the public exhibition of the draft EIS and the consideration of submissions made. A developer may appeal against the refusal of a development application, or against an approval subject to conditions which the developer finds unacceptable. Any person who has objected to "designated development" (i.e. those developments for which an EIS is required under part IV) may also bring an action in the Land and Environment Court on the merits of the application. The Land and Environment Court is a specialist tribunal established at the same time as the *Environmental Planning and Assessment Act* was enacted in 1979. In these appeals, the decision of the Land and Environment Court stands in the place of that of the consent authority, and is binding on all parties, including the Government.

The Act also incorporates an open standing provision, which applies to all parts of the Act. This provision allows "any person" to take action to remedy or restrain a breach of the Act, whether or not the right of that person has or may be infringed as a consequence of the breach. During the 1980s a relatively liberal legal aid system, specifically allowing for assistance in environmental matters, enabled the taking of various actions against both private and public sector developers. These actions have resulted in a jurisprudence of public interest law similar in some ways to that in the United States and Canada. However, legal aid in civil cases was severely curtailed in 1992; this development is likely to reduce the ability of community groups and environmental organisations to engage in public interest litigation in this area.

Other applicable legislation in New South Wales in the context of this case study is the *Pollution Control Act* 1970, the *Clean Waters Act* 1970 and the *Clean Air Act* 1961. These statutes have been under review for some time.

In NSW the pollution control functions of the State Government are undertaken by the Environment Protection Authority (EPA) under the *Protection of the Environment Administration Act* 1991. Until 1992, the relevant agency was the State Pollution Control Commission. Developments must satisfy the requirements of the *Clean Waters Act*, the *Clean Air Act* and the *Noise Control Act*, and must obtain the appropriate licences for industrial operations. Standards set for emissions into the atmosphere by the Tomago plant have been largely based on those prescribed by the regulations made under the *Clean Air Act*. However, with the expansion of the plant during the early 1990s, the EPA has been engaged in negotiations to impose tighter standards across a comprehensive range of emissions.

Federal and State Law Applicable to the Case Study

The Tomago Aluminium proposal involved financial agreements subject to the Commonwealth Government's foreign investment policy, so the EIS was subject to both State and Federal approval. Where Federal and State legislation both apply, it is usual for the Federal and State governments to enter into consultations in order to determine the way in which the environmental impact assessment process is to be carried out. These consultations are generally guided by a Memorandum of Understanding (MOU) previously drawn up between the relevant governments. Such a MOU was signed between the NSW and Commonwealth Governments in 1983. In this case study the EIS conforms to both pieces of legislation.

The Hunter Regional Environmental Plan

Development in New South Wales is governed by local environmental (land use) plans, which generally must conform to regional environmental plans administered by the Department of Planning. Regional environmental plans can apply to large areas of land in order to deal more adequately with planning issues in a more coherent and integrated way. In the Hunter region, the expansion of industrial and mining activities and the consequent decline in the relative importance of agriculture has been an inevitable source of conflict. The conflict between the various land uses for agriculture, industry and tourism constitutes a complex planning issue, which was perceived to be best dealt with through regional planning. The Hunter Regional Environmental Plan, which was begun in the 1970s, became a statutory plan under the *Environmental Planning and Assessment Act* when that legislation came into force in 1980. The plan is meant to ensure that rational planning processes are applied over the region in relation to development activities. The plan, which was extensively revised in 1989, includes provisions relating to social and economic development, land use, natural resources, environment protection and the conservation of natural and recreational areas.

Description of the Development Proposal

In the early 1980s, Tomago Aluminium Company Pty Ltd constructed an aluminium smelter at Tomago, 12.5 km north of Newcastle. The smelter was designed with an annual capacity of 220,000 tonnes, requiring electricity power capacity of 380 MW, or 7% the total generating capacity of the State. Assuming that the base-load power stations typically operate at 60% of capacity, this indicated the need for extra generating capacity of 630 MW.

Agreements were entered into with the NSW Electricity Commission for the supply of electricity from the State grid. The extension of transmission lines to the Tomago site was the subject of a separate environmental impact assessment under NSW State legislation.

Aluminium smelting involves the electrolysis of alumina to produce aluminium and oxygen. Fluoride compounds are used in the electrolytic bath solutions that form the basis of the smelting process, resulting in the emission of gaseous fluoride and particulate fluoride compounds. Other residuals of concern are sulphur oxides.

To control these emissions it was proposed that pot hoods with collection efficiencies of 97.5% be installed. Collected emissions would then be treated by dry scrubbing. This involves the adsorption of gaseous fluorides onto alumina, and their consequent collection, along with particulate fluorides, by bag filters. Matter thus collected would then be fed back into the reduction cells. The dry scrubbing efficiency was expected to be 99.9%.

Analysis of the Environmental Effects of the Proposal

Sources of Environmental Impacts

The aim of the EIA undertaken for the Tomago aluminium smelter was to identify and explore the implications of environmental effects predicted from the construction and operation of the smelter (Croft and Associates 1980).

Identified effects included impacts on air and water quality; land use in the surrounding area; disposal of solid waste; and impacts on the social environment, such as the additional demands on housing, electricity supplies, community facilities, and sewerage. The threat posed to ambient air quality by fluoride emissions was of particular concern and became the focus of a special environmental study.

When a smelter is introduced, background concentrations of fluorides are supplemented by discharges from the plant. The main sources of fluorides are hydrogen fluoride from the smelter stacks and potline roof vents, fluoride absorbed on the surfaces of particulate emissions, and water soluble particulate emissions such as sodium fluoride. It was calculated that 28 kg of fluoride per tonne of aluminium would be emitted from the electrolytic cells. The scrubbing process would remove 99.9%, leaving a residual of 0.04 kg of fluoride per tonne of aluminium to be released into the atmosphere. An additional 0.7 kg per tonne of aluminium would be released in the form of fugitive emissions.

The Dispersion Model

Dispersion modelling was undertaken to predict the effects of the proposal on air quality. The aim of the pollution modelling exercise was to determine, with the highest possible degree of accuracy, the ground level concentration of gaseous and total fluoride both in the vicinity of the plant and for the region as a whole (Croft and Associates 1980). A modified Gaussian plume model was used.

The model incorporated dispersion parameters that varied with distance from the source and with meteorological conditions. Modifications were also made to account for downwash effects created by buildings. It was assumed that the small size of the particulate fluoride compounds would cause them to behave similarly to gaseous fluoride. It was also assumed that the plume would be non-reflective with respect to fluorides due to their high reactivity.

For the purpose of the analysis, a maximum emission rate of 1 kg of fluoride per tonne of aluminium was assumed. Meteorological data were obtained from the RAAF base at Williamtown, 13 km away.

The dispersion modelling was used to predict maximum monthly average ground level fluoride concentrations in the vicinity of the plant. Concentrations throughout the region were also modelled. Annual averages were derived by averaging the monthly concentrations.

Cumulative Impacts of Fluoride Emissions

The EIS considered the implications of fluoride emissions on the flora and fauna and rural land uses in the surrounding region and in the immediate vicinity of the plant. A regional perspective was incorporated to assess the cumulative impact of the smelter on ambient air quality.

Direct Effects of Fluoride Emissions on Flora and Fauna

It was expected that, in the long term, the fluoride emissions would alter the species composition of vegetation communities surrounding the plant. Effects would be most pronounced in the immediate vicinity of the plant, coinciding with the annual average isopleth for fluoride of 0.5 μg/cu m. Within the annual average isopleth of 0.3 μg/cu m sensitive crops including fruit and vegetables, and grazing animals and sensitive wildlife, were expected to be at risk. Within the 0.1 isopleth, barley and sorghum could be at risk.

For sulphur dioxide, it was predicted that the smelter would increase the ambient concentrations of sulphur dioxide by only about 4% throughout the region. Such an increase was not considered to have adverse environmental effects.

Monitoring Performance

With the plant now in actual operation, Tomago Aluminium has established a comprehensive and on-going sampling programme for vegetation and atmospheric parameters in an area several km around the plant to ensure compliance with environmental safety standards at all times.

A study examining the spatial and temporal patterns in the fluoride content of native vegetation up to a distance of 15 km from both the Tomago and Kurri Kurri smelters was undertaken between 1982 and 1985 (Bowen 1988). Vegetation damaged by fluoride was apparent only within a 0.5 km radius of Tomago, and elevated foliar fluoride concentrations were not found more than 1 km from the site. Though some sensitive species to the north-east and east were damaged during start-up, no vegetation death was apparent. These findings are quite different from the results of overseas experiences where extensive vegetation damage and death has occurred around long established smelters.

Economic Analysis of Tomago Aluminium Company's Proposal

Economic Analysis in the EIS

The Environmental Planning and Assessment Act in New South Wales requires that the social and economic effects of development activities should be taken into consideration when determining development applications under Part IV of the Act in relation to private sector activities, and under Part V of the Act in relation to public sector activities. The Environmental Impact Statement must include a justification of the proposed activity in terms of environmental, economic and social considerations. Details of these legal requirements can be found in James and Boer (1988).

The EIS for the Tomago smelter examined the predicted income and employment effects of the proposed smelter for the Hunter Region, for New South Wales and for Australia as a whole. Secondary income and employment effects were predicted using an input-output model.

"Cost-benefit analysis" was undertaken in the EIS to evaluate the desirability of the publicly-provided infrastructure from the point of view of the Government of New South Wales and its authorities.

Finally, the national economic impact was assessed with special reference to export earnings, with and without the aluminium smelter.

Direct Economic Effects of the Tomago Smelter

The original proposal indicated that the smelter would involve a capital investment of A$ 600 million during the construction phase. The greater proportion of this sum would be spent within Australia to finance initial design costs, buildings, plant and equipment and pollution abatement technology. An estimated 17% would be used to buy specialised equipment from other countries.

The proportion of funding from domestic sources was expected to approximate the proportion of Australian equity in the project. Significant interest and dividend payments would accrue to Australian nationals and this could be expected to contribute to the national economy.

With the smelter entering its operational phase, the main items of expenditure would include salaries and wages, tax payments to Commonwealth, State, and local governments, maintenance costs, the acquisition of raw materials, and power, transport and other indirect expenses.

Employment Effects in the Hunter Region, NSW and Australia

The direct employment figures for the operational phase of the smelter were based on similar overseas operations of the company. The direct employment figures for the construction phase were less certain as these would depend on the arrangements of contracting firms. Indirect employment occurs as a result of expanded sales for industries that benefit from construction and operation of the smelter. Induced employment results from the effects of increased consumer spending resulting from increased disposable income.

The impact of the smelter on regional employment opportunities was assessed using input-output analysis. The consultants made use of an existing transactions matrix for the region. This 25-sector matrix had been prepared by

Garlick (1979) from surveys conducted during the 1976-1977 financial year. Garner (cited in Croft and Associates 1980) expanded the model by incorporating an additional column and row to accommodate another proposed aluminium smelter at Lochinvar. The Lochinvar smelter was of the same size as the Tomago smelter and it was therefore considered acceptable to apply the Lochinvar multipliers to the analysis of the Tomago smelter. (The Lochinvar proposal, for various reasons, was abandoned.)

Both Type I and Type II multipliers for the Hunter region were used in the study, as shown in Table 10.1. These multipliers were used to predict the number of jobs that would result within the region from the direct, indirect, and induced effects of smelter construction and operation (see Table 10.2).

Type I multipliers account for the direct and indirect effects of a development. These multipliers can be used to calculate minimum likely employment (or income) effects of a development. Direct and indirect effects can be expected to occur immediately as the indirect effects represent the local industry support for the development.

Type II multipliers account for the direct, indirect, and induced effects of a development and represent an estimate of the upper bound of a development's economic impact. Type II effects take longer to materialise. In the study, a three year lag was adopted to reflect this fact. In addition, Type II multipliers were considered only for the operational phase and not for the construction phase, as the latter was considered to be of too short a duration to generate significant induced effects.

Table 10.1.

Employment Multipliers - Hunter Region

Construction phase:	Type I	1.64
	(direct and indirect)	
Operational phase:	Type I	1.75
	(direct and indirect)	
	Type II	2.75
	(direct, indirect and induced)	

Source: Croft and Associates (1980)

The employment effects of the smelter in the State of New South Wales and Australia as a whole were also calculated using multiplier analysis. In the absence of input-output tables for New South Wales, employment multipliers based on those developed for Queensland were adopted. Employment effects on a national scale were estimated using the latest Australian input-output tables; at the time, for the year 1968-1969. The multipliers were larger for the Australian economy as a whole than for the Hunter Region; expenditure "leakages" tend to be smaller the broader are the economic boundaries of analysis. For reasons of brevity, the results of these computations are not reported here.

Table 10.2.

Estimated Total New Employment within the Hunter Region as a Result of the Proposed Development.

	1980	1981	1982	1983	1984	1985	1986	1987	1988 onwards
	(1)	(2)	(3)	(4)	(5)	(6)	(7)	(8)	(9)
Direct Construction Workforce	50	1,200	1,500	1,500	1,200	500	-	-	-
Direct Operational Workforce	-	90	120	560	610	800	800	800	800
Indirect Employment (Construction)	-	770	960	960	770	320	-	-	-
Indirect Employment (Operational)	-	-	-	420	460	600	600	600	600
Induced Employment (Operational)	-	-	-	-	-	-	380	470	800
Total New Employment	50	2,060	2,580	3,440	3,040	2,220	1,780	1,870	2,200
Labour Force Projections[a]	148,900	150,300	152,000	153,700	155,500	157,200	159,000	160,800	162,500

Notes:
1. These figures are order of magnitude figures only.
2. A delay of 3 years for induced employment figures has been adopted.

[a] Labour force projections as used in the Hunter Regional Plan.

Source: Croft and Associates (1980)

Income Effects in the Hunter Region and Australia

The direct income effects of the smelter consist of wages and salaries paid to employees. This was estimated to be A$15 million per year (in 1980 dollars) when the plant would be fully operational. The direct wages paid during the construction phase were expected to total A$60 million.

Indirect income effects were estimated using income multipliers. These were estimated separately for the construction and operational phases of the plant. Multipliers for the Hunter Region were derived from the study by Garner, while for Australia as a whole the National Accounts Input-Output Tables 1968-1969 were used (see Table 10.3).

Table 10.3.

Income Multipliers - Hunter Region and Australia

	Construction Phase	Operational Phase	
	Type I	Type I	Type II
Hunter Region	1.43	1.49	2.15
Australia	1.87	2.39	3.38

Source: Croft and Associates (1980)

It was estimated that during the construction of the smelter the direct wages bill of A$60 million would generate an expected A$52 million of indirect income, and that approximately A$26 million of this amount would accrue to individuals in the Hunter Region.

In the operational phase, the A$15 million wage and salary bill was estimated to generate between A$21 million and A$36 million in additional income throughout Australia as a result of multiplier effects. Between A$7 million and A$17 million of this amount was estimated to accrue to residents of the Hunter Region.

Fiscal Analysis of Public Infrastructure

The construction and operation of the proposed Tomago smelter required the provision of public infrastructure. To address the question of whether the necessary government expenditure was justified, the consultants conducted a limited cost-benefit analysis of government outlays and revenues. The term "cost-benefit analysis" was loosely applied; it would have been more accurate to describe the analysis as a fiscal impact analysis or a cost-revenue analysis.

The main costs included the construction of an unloading berth for raw materials, the upgrading of 2 km of the Tomago Road, and additional road maintenance costs associated with the transport of raw materials and products between the smelter and the port of Newcastle.

The financial benefits identified consisted of tax revenues and other charges accruing to government bodies. In particular, financial benefits would accrue in the form of harbour and berthing charges, payroll tax, consumption tax, and income taxes returned to the State Government from the Commonwealth Government as part of the income tax sharing agreement, and revenues such as land tax and fuel tax.

A discount rate of 4% was used in this analysis. It was considered that this rate reflected the real cost of funds, given the prevailing government long-term borrowing rate of 12.6% and a long-term inflation rate of 8%.

The results of the analysis indicated that over a time horizon of 52 years (the assumed life of the aluminium smelter) a net benefit of A$ 60.4 million in present value terms would accrue to the Government of New South Wales and its authorities. The basic reasoning was that the Government of New South Wales would stand to gain this amount if it agreed to provide the necessary infrastructure.

A social benefit-cost analysis of infrastructure investment would have dealt with benefits and costs for the community as a whole, measuring the opportunity costs of resources required to provide the various government facilities, and the social benefits that would be generated. Transfer payments such as taxes would not have been relevant in such an analysis.

Balance of Payments Considerations

The EIS included a small section on "the national economic impact" which discussed the economic benefits of processing bauxite before export. This was shown to increase Australia's export earnings as well as provide additional employment and income.

It was explained that the production of one tonne of aluminium at Tomago would require an input of two tonnes of alumina. The production of alumina, in turn, requires four tonnes of bauxite. To quote the EIS, "at prices of $1550, $217 and $12 per tonne respectively, exported aluminium metal is 3.5 times more valuable than the equivalent amount of export alumina and 30 times more valuable than the equivalent amount of export bauxite." Furthermore, "in employment, the smelting of aluminium metal directly provides 2.7 times as many jobs as bauxite mining and export, for the equivalent output."

Also assessed was the annual export value of the raw material inputs to aluminium smelting at Tomago. In the absence of the smelter, inputs such as alumina and coal could be exported, whereas petroleum coke, another essential input, would not have to be imported. The export revenue from aluminium production compared favourably with this opportunity cost.

The economic significance of these export earnings and import savings that would be earned in the absence of the smelter was not explained in the EIS. In a social benefit-cost analysis, the calculations would be irrelevant.

A Systems Analysis Model of the Hunter Region

Rationale for a Pro-active Approach

The Tomago Aluminium Company EIS is an example of how, in a relatively simple case, the cumulative impacts of a development proposal may be assessed. The economic impact analysis was extensive. Cumulative environmental impacts, however, were not assessed in terms of regional production linkages.

In a region containing much industrial activity, such as the Hunter, and with further development expected in the future, a case-by-case approach to cumulative impact assessment is inadequate. A comprehensive assessment of potential impacts from all existing and planned activities is required to avoid exceeding the region's assimilative capacity. To achieve cost-effective control of pollution, area-wide anticipatory planning should be undertaken. If this is not done, late arrivals may be forced to bear a disproportionate share of pollution control costs, leading to an excessive cost burden and a waste of economic resources.

It would be unfair and economically inefficient to insist that, each time a new development proposal is put forward, the proponent should undertake a full-scale regional economic-environmental modelling assessment. Such an approach could lead, among other things, to unnecessary duplication of research and inconsistencies in modelling results.

Cumulative effects should be assessed and managed within a broad environmental planning framework. CEARC (1986) assert that, due to the pervasiveness of cumulative effects, they should be managed on an area-wide planning basis rather than on a project basis. A regional planning body can adopt a synoptic view, develop expertise specific to the region and conduct ongoing studies and assessment. Systems models can be constructed to support individual EIAs as well as facilitate regional development planning.

Optimal use could be made of the assimilative capacity of the environment, and fragile areas could be protected from undue development pressure. Regional guidelines can simplify the task of project-specific EIA, by indicating which proposals should be rejected or accepted on regional criteria. The task of the traditional EIA is thus simplified, and the EIA team is free to examine other potential environmental issues and collect supplementary information on key cumulative impacts.

An Integrated Regional Systems Model

A comprehensive assessment of the effects of development, including aluminium smelting, on air quality in the Hunter region was undertaken by a multidisciplinary team on a research project supported by the Australian National Energy Research Development and Demonstration Council (James et al. 1983). A two-stage approach was adopted. First, an input-output model was used to determine production levels by all industries in the region under alternative development scenarios. The input-output model was based on tables compiled by Garlick (1979). A key to the sectors is provided in Table 10.4.

Table 10.4.

Key to Input-Output sectors

1 Raw Coal Production	14 Glass, Clay, Nonmetallic Minerals
2 Coal Washeries	15 Wood, Wood Products, Furniture
3 Electricity Generation	16 Paper, Paper Products, Printing
4 Aluminium Smelting	17 Other Manufacturing
5 Rural	18 Electricity, Gas, Water Utilities
6 Other Mining	19 Building and Construction
7 Food, Beverages, Tobacco	20 Wholesale, Retail Trade
8 Textiles, Clothing, Footwear	21 Transport, Storage, Communication
9 Basic Iron and Steel	22 Finance, Business, Professional Services
10 Other Nonferrous Basic Metals	23 Public Authorities, Defence
11 Fabricated Metal Products	24 Community, Other Services
12 Transport Equipment	25 Dwelling Rent
13 Chemicals and Petroleum Products	

Source: James and Chambers (1986)

The model was modified by using physical units to measure the outputs of key sectors such as aluminium smelting, electricity generation and coal production. Second, emissions resulting from the economic pattern were determined, and a dispersion model used to map resultant ambient concentrations of the main air pollutants (sulphur oxides and fluorides).

In a study conducted by James and Chambers (1986) three development scenarios (A, B and C) were explored with the model. Scenario A was based on projections of final demands to the year 1990-91 prepared by the Hunter Valley Research Foundation, based on market surveys and industry plans. Scenario B assumed greater capacity utilisation within the power industry, with base load stations operating at 65% of full capacity, intermediate load stations at 50% and peak load stations at 30%. Scenario B was expected to occur in the 1990s. Scenario C explored short-term conditions, with all stations operating at full capacity. Scenario C would occur only under exceptional conditions, and only during a time of peak demand for electricity.

Estimation of Sulphur Oxide and Fluoride Emissions

In the modelling work, emissions of sulphur oxides and fluorides were estimated for each economic sector. Use of physical units for key sectors permitted straightforward application of discharge coefficients based on engineering data. Coefficients for the power stations were based on US Environmental Protection Agency combustion emission data. Two different discharge coefficients were used for the two aluminium smelters in the Hunter, reflecting the different pollution control technologies used in the two plants. These coefficients were obtained from a study published by the NSW State Pollution Control Commission (1980) and from the EIS compiled by Croft and Associates (1980).

Discharge coefficients for the remaining sectors of the model were obtained using an existing emissions inventory of the Hunter Region. The complete set of discharge coefficients is shown in Table 10.5.

Emissions under Scenario A are shown in Table 10.6. It is worth noting that electric power stations were predicted to discharge three times as much fluoride, in total, as the aluminium smelters. Indeed, discharges from the smelters were predicted to decline from 670 t/yr in 1976-77 to 404 t/yr, even though aluminium output increases from 45,000 t/yr to 355,000 t/yr over the planning period. This is a consequence of strict emission standards.

Table 10.5.
Sulphur Oxide and Fluoride Discharge Coefficients
for Input-Output Sectors (kg/$'000)

Sector	SO_x direct	SO_x cumulated	F direct	F cumulated
1[a]	-	25.7	-	0.28
2[a]	-	35.8	-	0.39
3[b]	3,624.0	3,637.0	35.76	35.90
4[c]	26.7	78.9	1.14	1.66
5	-	5.1	-	0.05
6	0.7	4.3	-	0.04
7	1.2	4.8	0.01	0.04
8	4.5	8.8	0.02	0.06
9	12.7	20.8	0.15	0.23
10	45.6	47.3	-	0.02
11	1.5	5.1	-	0.04
12	-	2.1	-	0.02
13	50.3	52.1	0.42	0.43
14	3.9	6.5	0.60	0.67
15	-	1.5	0.20	0.04
16	-	0.8	-	0.01
17	-	2.9	-	0.05
18	-	52.0	-	0.52
19	-	1.2	-	0.03
20	-	1.3	-	0.02
21	-	2.2	-	0.02
22	-	0.2	-	-
23	-	3.4	-	0.40
24	0.1	0.9	-	0.01
25	-	0.5	-	0.01

[a] kg/'000 tonnes

[b] kg/GWh

[c] kg/tonne

- Indicates zero or insignificant value

Source: James and Chambers (1986)

Table 10.6.
Final Demands, Sectoral Outputs, Sulphur Oxide and Fluoride Discharges for Base Year and Scenario A.

Sector	Sectoral Outputs 1976-77 $m	SO_x Discharges 1976-77 tonnes	Fluoride Discharges 1976-77 tonnes	Sectoral Outputs 1990-91 $m	SO_x Discharges 1990-91 tonnes	Fluoride Discharges 1990-91 tonnes
(1)	(2)	(3)	(4)	(5)	(6)	(7)
1	27,408[a]	-	-	49,070[a]	-	-
2	10,754[a]	-	-	22,156[a]	-	-
3	22,227[b]	80,550	795	33,976[b]	123,131	1,215
4	45,000[c]	1,200	670	355,000[c]	9,467	404
5	109.2	-	-	117.3	-	-
6	45.3	33	-	45.6	33	-
7	200.6	240	1	224.4	269	2
8	66.1	296	1	75.7	340	2
9	616.4	7,817	89	511.1	6,481	74
10	61.8	2,822	-	97.6	4,453	-
11	152.8	228	-	260.1	388	-
12	81.3	2	-	153.0	3	-
13	151.8	7,627	63	55.1	2,769	23
14	70.5	276	42	76.3	299	46
15	49.6	-	1	52.1	-	1
16	27.4	-	-	29.0	-	-
17	106.0	-	-	187.9	-	-
18	134.3	-	-	177.7	-	-
19	623.7	-	-	780.9	-	-
20	1,041.8	3	-	1,234.1	4	-
21	410.5	-	-	484.4	-	-
22	594.9	-	-	737.5	-	-
23	174.8	-	-	186.7	-	-
24	653.8	65	1	1,239.6	124	1
25	51.6	-	-	95.6	-	-
Total		101,164	1,664		147,765	1,768

[a] '000 tonnes; [b] GWh; [c] Tonnes; - Indicates zero or insignificant value;

Source: James and Chambers (1986).

The Dispersion Model

Ambient pollutant concentrations were predicted by disaggregating the sectoral emissions into discharges for individual plants, plotting these sources on a grid, and applying an air dispersion model. A modified Gaussian plume model established by Dennis (1978) was used in the study. Area concentrations were converted to concentrations at receptor points on the map grid. The model's reliability was tested by comparisons of predicted and monitored sulphur oxide and fluoride concentrations for 1979-80.

The model was used to predict ambient concentrations of sulphur oxides and fluorides for yearly and monthly averages in 1976-77 and 1990-91. It predicted that fluorides were not likely to pose serious problems; maximum monthly average concentrations set by the SPCC (0.1 µg/cu m) would not be exceeded in the main agricultural areas of the Hunter. Even though power stations discharge a greater load of fluoride than aluminium smelters, their tall stacks result in effective dispersion. The contribution by power stations to ambient concentrations at ground level is minimal. Fluorides from smelters are dispersed largely in buffer zones around the smelters themselves. Further study was warranted on short-term concentrations, especially in summer when prevailing winds carry fluorides towards vineyards in the region.

Monthly average concentrations of sulphur oxides predicted by the study fell within the Australian Health and Medical Research Council and World Health Organization's recommended annual levels of 60 µg/cu m. However, there was a prospect that the WHO standard would be lowered to 40 µg/cu m. The model predicted concentrations around this level in the most densely populated parts of the region. Short term peak concentrations of sulphur oxides were likely to exceed WHO and US standards in some areas by the early 1990s.

Economic Assessment of Alternative Scenarios

Models such as the NERDDC model are capable of simulating the impact of different development plans, environmental management strategies and pollution control technologies, especially for major industrial developments such as aluminium smelting. James et al. (1985) have suggested a simple two-criteria technique for rejection or acceptance of an alternative: to be accepted the alternative (i) must be economic and (ii) must achieve the desired environmental result. Strategies that fail to achieve the desired result can usually be improved, but normally only at an increased cost. Thus, the economic criterion must be reapplied before the strategy is accepted.

Another evaluation approach is full benefit-cost analysis. This involves extensive data requirements and modelling work. There could also be difficulties placing economic values on all environmental effects of predicted ambient concentrations of sulphur oxides and fluorides.

Cost-effectiveness analysis can be undertaken where environmental damage costs or protection benefits are difficult to estimate. Under this approach, strategies are evaluated in terms of their ability to meet environmental standards at minimum cost. Given that ambient concentration standards have been established for major air pollutants in the Hunter, simulation modelling accompanied by cost-effectiveness analysis would appear to be the most realistic way to conduct comprehensive economic and environmental assessments for the region.

References

Bates, G. (1987) *Environmental Law in Australia*, Butterworths, Sydney.

Bowen, S.E. (1988) 'Spatial and Temporal Patterns in the Fluoride Content of Vegetation Around Two Aluminium Smelters in the Hunter Valley, New South Wales', *The Science of the Total Environment*, 68:97-111.

Canadian Environmental Assessment Research Council and US National Research Council Board on Basic Biology (1986) 'Proceedings on the Workshop on Cumulative Environmental Effects: A Binational Perspective'.

Croft, J.B., and Associates (1980). *Environmental Impact Statement for an Aluminium Smelter at Tomago, NSW, Vols.1, 2 and 3*, Newcastle.

Dennis, R.L. (1978) 'The Smeared Concentration Approximation Method: A Simplified Air Pollution Dispersion Methodology for Regional Analysis', International Institute for Applied Systems Analysis, Laxenburg, Austria.

Garlick, S. (1979) 'New Input-Output Tables for the Hunter Region', Hunter Valley Research Foundation, Newcastle.

James, D. and B.W. Boer (1988) 'Application of Economic Techniques in Environmental Impact Assessment: Preliminary Report', Australian Environment Council, Canberra.

James, D., J. Chambers, A. Gilbert and H. Wright (1983) 'An Integrated Energy-Economic-Environmental Model for Australia', report prepared for the National Energy Research, Development and Demonstration Council, NERDDP Project No 80/0303, Canberra.

James, D. and J. Chambers (1986) 'Managing Environmental Impacts of Energy Development: Air Quality in the Hunter Region, New South Wales, Australia', *Environmental Management*, 10,3:421-430.

James, D.E., J.A. Chambers, J.D. Kalma, and H.A. Bridgman (1985) 'Air Quality Prediction in Urban and Semi-Urban Regions With Generalized Input-Output Analysis: The Hunter Region, Australia', *Urban Ecology*, 9:25-44.

NSW State Pollution Control Commission (1980) 'Pollution Control in the Hunter Valley with Particular Reference to Aluminium Smelting', Sydney.

Environment & Management

1. O. Kuik and H. Verbruggen (eds.): *In Search of Indicators of Sustainable Development.* 1991 ISBN 0-7923-1249-X
2. R. Janssen: *Multiobjective Decision Support for Environmental Management.* 1992 ISBN 0-7923-1908-7
3. R. Janssen and M. van Herwijnen: *DEFINITE. A System to Support DEcisions on a FINITE Set of Alternatives.* Software with User Manual. 1994 ISBN 0-7923-2696-2
4. D. James: *The Application of Economic Techniques in Environmental Impact Assessment.* 1994 ISBN 0-7923-2721-7

KLUWER ACADEMIC PUBLISHERS – DORDRECHT / BOSTON / LONDON